Mastering Shiny

*Build Interactive Apps, Reports,
and Dashboards Powered by R*

Hadley Wickham

Beijing · Boston · Farnham · Sebastopol · Tokyo

Mastering Shiny

by Hadley Wickham

Published by O'Reilly Media, Inc., 1005 Gravenstein Highway North, Sebastopol, CA 95472.

O'Reilly books may be purchased for educational, business, or sales promotional use. Online editions are also available for most titles (*http://oreilly.com*). For more information, contact our corporate/institutional sales department: 800-998-9938 or *corporate@oreilly.com*.

Acquisitions Editor: Jessica Haberman
Development Editor: Melissa Potter
Production Editor: Christopher Faucher
Copyeditor: nSight, Inc.
Proofreader: Piper Editorial Consulting, LLC

Indexer: Judith McConville
Interior Designer: David Futato
Cover Designer: Karen Montgomery
Illustrator: Kate Dullea

May 2021:   First Edition

Revision History for the First Edition
2021-04-29:   First Release

See *http://oreilly.com/catalog/errata.csp?isbn=9781492047384* for release details.

978-1-492-04738-4

[LSI]

Table of Contents

Preface. xiii

Part I.   Getting Started

1. Your First Shiny App.   3

Introduction	3
Create App Directory and File	3
Running and Stopping	4
Adding UI Controls	6
Adding Behavior	7
Reducing Duplication with Reactive Expressions	8
Summary	9
Exercises	10

2. Basic UI.   15

Introduction	15
Inputs	15
Common Structure	16
Free Text	16
Numeric Inputs	17
Dates	18
Limited Choices	18
File Uploads	20
Action Buttons	20
Exercises	21
Outputs	22
Text	22

Tables   24
Plots   25
Downloads   25
Exercises   26
Summary   26

3. Basic Reactivity... 27

Introduction   27
The Server Function   27
Input   28
Output   29
Reactive Programming   30
Imperative Versus Declarative Programming   31
Laziness   32
The Reactive Graph   33
Reactive Expressions   33
Execution Order   34
Exercises   35
Reactive Expressions   36
The Motivation   36
The App   38
The Reactive Graph   40
Simplifying the Graph   41
Why Do We Need Reactive Expressions?   43
Controlling Timing of Evaluation   44
Timed Invalidation   45
On Click   46
Observers   49
Summary   50

4. Case Study: ER Injuries... 51

Introduction   51
The Data   51
Exploration   53
Prototype   57
Polish Tables   60
Rate Versus Count   61
Narrative   63
Exercises   64
Summary   64

Part II. Shiny in Action

5. Workflow.. **67**

Development Workflow   67

    Creating the App   68

    Seeing Your Changes   69

    Controlling the View   70

Debugging   70

    Reading Tracebacks   71

    Tracebacks in Shiny   72

    The Interactive Debugger   74

    Case Study   75

    Debugging Reactivity   79

Getting Help   80

    Reprex Basics   81

    Making a Reprex   81

    Making a Minimal Reprex   82

    Case Study   83

Summary   87

6. Layout, Themes, HTML.. **89**

Introduction   89

Single-Page Layouts   89

    Page Functions   90

    Page with Sidebar   91

    Multirow   93

    Exercises   94

Multipage Layouts   94

    Tabsets   94

    Navlists and Navbars   96

Bootstrap   97

Themes   98

    Getting Started   98

    Shiny Themes   99

    Plot Themes   100

    Exercises   101

Under the Hood   101

Summary   103

7. Graphics... **105**

Interactivity   105

    Basics   105

Clicking	107
Other Point Events	109
Brushing	109
Modifying the Plot	111
Interactivity Limitations	115
Dynamic Height and Width	115
Images	116
Summary	118

8. User Feedback. **119**

Validation	119
Validating Input	120
Canceling Execution with req()	121
req() and Validation	124
Validate Output	125
Notifications	126
Transient Notification	127
Removing on Completion	128
Progressive Updates	129
Progress Bars	129
Shiny	130
Waiter	132
Spinners	133
Confirming and Undoing	136
Explicit Confirmation	136
Undoing an Action	137
Trash	139
Summary	139

9. Uploads and Downloads. **141**

Upload	141
UI	141
Server	142
Uploading Data	143
Download	144
Basics	144
Downloading Data	145
Downloading Reports	146
Case Study	149
Exercises	151
Summary	152

10. Dynamic UI. **153**
    Updating Inputs   153
      Simple Uses   155
      Hierarchical Select Boxes   156
      Freezing Reactive Inputs   158
      Circular References   160
      Interrelated Inputs   160
      Exercises   161
    Dynamic Visibility   162
      Conditional UI   163
      Wizard Interface   165
      Exercises   166
    Creating UI with Code   166
      Getting Started   167
      Multiple Controls   168
      Dynamic Filtering   171
      Dialog Boxes   175
      Exercises   176
    Summary   177

11. Bookmarking. **179**
    Basic Idea   179
      Updating the URL   182
      Storing Richer State   182
    Bookmarking Challenges   183
    Exercises   184
    Summary   184

12. Tidy Evaluation. **185**
    Motivation   185
    Data-Masking   187
      Getting Started   187
      Example: ggplot2   189
      Example: dplyr   191
      User-Supplied Data   193
      Why Not Use Base R?   194
    Tidy-Selection   195
      Indirection   195
      Tidy-Selection and Data-Masking   196
    parse() and eval()   197
    Summary   197

Part III.   Mastering Reactivity

13. Why Reactivity?. **201**

Introduction   201

Why Do We Need Reactive Programming?   202

  Why Can't You Use Variables?   202

  What About Functions?   202

  Event-Driven Programming   203

  Reactive Programming   204

A Brief History of Reactive Programming   206

Summary   207

14. The Reactive Graph. **209**

Introduction   209

A Step-by-Step Tour of Reactive Execution   209

A Session Begins   211

  Execution Begins   211

  Reading a Reactive Expression   212

  Reading an Input   213

  Reactive Expression Completes   213

  Output Completes   214

  The Next Output Executes   214

  Execution Completes, Outputs Flushed   214

An Input Changes   215

  Invalidating the Inputs   215

  Notifying Dependencies   216

  Removing Relationships   216

  Re-execution   217

  Exercises   217

Dynamism   218

The Reactlog Package   220

Summary   221

15. Reactive Building Blocks. **223**

Reactive Values   223

  Exercises   224

Reactive Expressions   225

  Errors   225

  on.exit()   226

  Exercises   226

Observers and Outputs   226

Isolating Code   228 isolate() 228

observeEvent() and eventReactive() 229

Exercises 230

Timed Invalidation 230

Polling 231

Long-Running Reactives 231

Timer Accuracy 232

Exercises 233

Summary 233

16. Escaping the Graph. **235**

Introduction 235

What Doesn't the Reactive Graph Capture? 235

Case Studies 237

One Output Modified by Multiple Inputs 237

Accumulating Inputs 238

Pausing Animations 239

Exercises 240

Antipatterns 240

Summary 242

Part IV.   Best Practices

17. General Guidelines. **245**

Introduction 245

Code Organization 246

Testing 247

Dependency Management 247

Source Code Management 248

Continuous Integration/Deployment 249

Code Reviews 249

Summary 250

18. Functions. **251**

File Organization 252

UI Functions 252

Other Applications 253

Functional Programming 254

UI as Data 254

Server Functions 255

Reading Uploaded Data 255

Internal Functions	256
Summary	257

19. Shiny Modules **259**

Motivation	259
Module Basics	261
Module UI	262
Module Server	262
Updated App	263
Namespacing	264
Naming Conventions	265
Exercises	265
Inputs and Outputs	266
Getting Started: UI Input and Server Output	267
Case Study: Selecting a Numeric Variable	268
Server Inputs	269
Modules Inside of Modules	270
Case Study: Histogram	271
Multiple Outputs	272
Exercises	274
Case Studies	275
Limited Selection and Other	275
Wizard	278
Dynamic UI	282
Single Object Modules	284
Summary	286

20. Packages **287**

Converting an Existing App	288
Single File	289
Module Files	290
A Package	291
Benefits	292
Workflow	292
Sharing	293
Extra Steps	294
Deploying Your App-Package	294
R CMD check	294
Summary	296

21. Testing **297**

Testing Functions	298

Basic Structure	298
Basic Workflow	299
Key Expectations	300
User Interface Functions	302
Workflow	304
Code Coverage	304
Keyboard Shortcuts	304
Workflow Summary	305
Testing Reactivity	305
Modules	307
Limitations	309
Testing JavaScript	309
Basic Operation	310
Case Study	311
Testing Visuals	313
Philosophy	314
When Should You Write Tests?	314
Summary	315

22. Security.. 317

Data	318
Compute Resources	319

23. Performance.. 323

Dining at Restaurant Shiny	324
Benchmark	325
Recording	325
Replay	326
Analysis	327
Profiling	328
The Flame Graph	329
Profiling R Code	331
Profiling a Shiny App	332
Limitations	333
Improve Performance	333
Caching	334
Basics	334
Caching a Reactive	335
Caching Plots	336
Cache Key	337
Cache Scope	338
Other Optimizations	338

Schedule Data Munging   338
Manage User Expectations   339
Summary   340

Index. 341

Preface

What Is Shiny?

If you've never used Shiny before, welcome! Shiny is an R package that allows you to easily create rich, interactive web apps. Shiny allows you to take your work in R and expose it via a web browser so that anyone can use it. Shiny makes you look awesome by making it easy to produce polished web apps with a minimum amount of pain.

In the past, creating web apps was hard for most R users because:

- You need a deep knowledge of web technologies like HTML, CSS, and JavaScript.
- Making complex interactive apps requires careful analysis of interaction flows to make sure that when an input changes, only the related outputs are updated.

Shiny makes it significantly easier for the R programmer to create web apps by:

- Providing a carefully curated set of user interface (UI for short) functions that generate the HTML, CSS, and JavaScript needed for common tasks. This means that you don't need to know the details of HTML/CSS/JavaScript until you want to go beyond the basics that Shiny provides for you.
- Introducing a new style of programming called *reactive programming*, which automatically tracks the dependencies of pieces of code. This means that whenever an input changes, Shiny can automatically figure out how to do the smallest amount of work to update all the related outputs.

People use Shiny to:

- Create dashboards that track important high-level performance indicators while facilitating drill-down into metrics that need more investigation.
- Replace hundreds of pages of PDFs with interactive apps that allow the user to jump to the exact slice of the results that they care about.

- Communicate complex models to a nontechnical audience with informative visualizations and interactive sensitivity analysis.

- Provide self-service data analysis for common workflows, replacing email requests with a Shiny app that allows people to upload their own data and perform standard analyses. You can make sophisticated R analyses available to users with no programming skills.

- Create interactive demos for teaching statistics and data science concepts that allow learners to tweak inputs and observe the downstream effects of those changes in an analysis.

In short, Shiny gives you the ability to pass on some of your R superpowers to anyone who can use the web.

Who Should Read This Book?

This book is aimed at two main audiences:

- R users who are interested in learning about Shiny in order to turn their analyses into interactive web apps. To get the most out of this book, you should be comfortable using R to do data analysis and should have written at least a few functions.

- Existing Shiny users who want to improve their knowledge of the theory underlying Shiny in order to write higher-quality apps faster and more easily. You should find this book particularly helpful if your apps are starting to get bigger and you're starting to have problems managing the complexity.

What Will You Learn?

The book is divided into four parts:

1. In Part I, you'll learn the basics of Shiny so you can get up and running as quickly as possible. You'll learn about the basics of app structure, useful UI components, and the foundations of reactive programming.

2. Part II builds on the basics to help you solve common problems, including giving feedback to the user, uploading and downloading data, generating UI with code, reducing code duplication, and using Shiny to program the tidyverse.

3. In Part III, you'll go deep into the theory and practice of reactive programming, the programming paradigm that underlies Shiny. If you're an existing Shiny user, you'll get the most value out of this chapter as it will give you a solid theoretical underpinning that will allow you to create new tools specifically tailored for your problems.

4. Finally, in Part IV we'll finish up with a survey of useful techniques for making your Shiny apps work well in production. You'll learn how to decompose complex apps into functions and modules, how to use packages to organize your code, how to test your code to ensure it's correct, and how to measure and improve performance.

What Won't You Learn?

The focus of this book is making effective Shiny apps and understanding the underlying theory of reactivity. I'll do my best to showcase best practices for data science, R programming, and software engineering, but you'll need other references to master these important skills. If you enjoy my writing in this book, you might enjoy my other books on these topics: *R for Data Science* (*http://r4ds.had.co.nz*), *Advanced R* (*http://adv-r.hadley.nz*), and *R Packages* (*http://r-pkgs.org*).

There are also a number of important topics specific to Shiny that I don't cover:

- This book only covers the built-in user interface toolkit. This doesn't provide the sexiest possible design, but it's simple to learn and gets you a long way. If you have additional needs (or just get bored with the defaults), there are a number of other packages that provide alternative frontends. See "Bootstrap" on page 97 for more details.

- Deployment of Shiny apps. Putting Shiny "into production" is outside the scope of this book because it hugely varies from company to company, and much of it is unrelated to R (the majority of challenges tend to be cultural or organizational, not technical). If you're new to Shiny in production, I recommend starting with Joe Cheng's 2019 rstudio::conf keynote (*https://oreil.ly/XNCRf*). That will give you the lay of the land, discussing broadly what putting Shiny into production entails and how to overcome some of the challenges that you're likely to face. Once you've done that, see the RStudio Connect website (*https://oreil.ly/FdrYc*) to learn about RStudio's product for deploying apps within your company, and the Shiny website (*https://oreil.ly/z8kRP*) for other common deployment scenarios.

Prerequisites

Before we continue, make sure you have all the software you need for this book:

R

> If you don't have R installed already, you may be reading the wrong book; I assume a basic familiarity with R throughout this book. If you'd like to learn how to use R, I'd recommend my *R for Data Science* (*https://r4ds.had.co.nz*), which is designed to get you up and running with R with a minimum of fuss.

RStudio

RStudio is a free and open source integrated development environment (IDE) for R. While you can write and use Shiny apps with any R environment (including R GUI and ESS (*http://ess.r-project.org*)), RStudio has some nice features specifically for authoring, debugging, and deploying Shiny apps. We recommend downloading RStudio Desktop (*https://oreil.ly/aUoYe*) and giving it a try, but it's not required to be successful with Shiny or with this book.

R packages

This book uses a bunch of R packages. You can install them all at once by running:

```
install.packages(c(
  "gapminder", "ggforce", "gh", "globals", "openintro", "profvis",
  "RSQLite", "shiny", "shinycssloaders", "shinyFeedback",
  "shinythemes", "testthat", "thematic", "tidyverse", "vroom",
  "waiter", "xml2", "zeallot"
))
```

If you've downloaded Shiny in the past, make sure that you have at least version 1.6.0.

Conventions Used in This Book

The following typographical conventions are used in this book:

Italic

Indicates new terms, URLs, email addresses, filenames, and file extensions.

`Constant width`

Used for program listings, as well as within paragraphs to refer to program elements such as variable or function names, databases, data types, environment variables, statements, and keywords.

`Constant width bold`

Shows commands or other text that should be typed literally by the user.

`Constant width italic`

Shows text that should be replaced with user-supplied values or by values determined by context.

This element signifies a tip or suggestion.

 This element signifies a general note.

Using Code Examples

Supplemental material (code examples, exercises, etc.) is available for download at *https://mastering-shiny.org*. The code samples in this book are licensed under the MIT License (*https://www.mit.edu/~amini/LICENSE.md*).

If you have a technical question or a problem using the code examples, please send email to *bookquestions@oreilly.com*.

This book is here to help you get your job done. In general, if example code is offered with this book, you may use it in your programs and documentation. You do not need to contact us for permission unless you're reproducing a significant portion of the code. For example, writing a program that uses several chunks of code from this book does not require permission. Selling or distributing examples from O'Reilly books does require permission. Answering a question by citing this book and quoting example code does not require permission. Incorporating a significant amount of example code from this book into your product's documentation does require permission.

We appreciate, but generally do not require, attribution. An attribution usually includes the title, author, publisher, and ISBN. For example: "*Mastering Shiny* by Hadley Wickham (O'Reilly). Copyright 2021 Hadley Wickham, 978-1-492-04738-4."

If you feel your use of code examples falls outside fair use or the permission given above, feel free to contact us at *permissions@oreilly.com*.

O'Reilly Online Learning

 For more than 40 years, *O'Reilly Media* has provided technology and business training, knowledge, and insight to help companies succeed.

Our unique network of experts and innovators share their knowledge and expertise through books, articles, and our online learning platform. O'Reilly's online learning platform gives you on-demand access to live training courses, in-depth learning paths, interactive coding environments, and a vast collection of text and video from O'Reilly and 200+ other publishers. For more information, visit *http://oreilly.com*.

How to Contact Us

Please address comments and questions concerning this book to the publisher:

O'Reilly Media, Inc.
1005 Gravenstein Highway North
Sebastopol, CA 95472
800-998-9938 (in the United States or Canada)
707-829-0515 (international or local)
707-829-0104 (fax)

We have a web page for this book, where we list errata, examples, and any additional information. You can access this page at *https://oreil.ly/mastering-shiny*.

Email *bookquestions@oreilly.com* to comment or ask technical questions about this book.

For news and information about our books and courses, visit *http://oreilly.com*.

Find us on Facebook: *http://facebook.com/oreilly*

Follow us on Twitter: *http://twitter.com/oreillymedia*

Watch us on YouTube: *http://youtube.com/oreillymedia*

Acknowledgments

This book was written in the open, and chapters were advertised on Twitter when complete. It is truly a community effort: many people read drafts, fixed typos, suggested improvements, and contributed content. Without those contributors, the book wouldn't be nearly as good as it is, and I'm deeply grateful for their help.

A big thank-you to all 83 people who contributed specific improvements via GitHub pull requests (in alphabetical order by username): Adam Pearce (@1wheel), Adi Sarid (@adisarid), Alexandros Melemenidis (@alex-m-ffm), Anton Klåvus (@antonvsdata), Betsy Rosalen (@betsyrosalen), Michael Beigelmacher (@brooklynbagel), Bryan Smith (@BSCowboy), c1au6io_hh (@c1au6i0), @canovasjm, Chris Beeley (@ChrisBeeley), @chsafouane, Chuliang Xiao (@ChuliangXiao), Conor Neilson (@condwanaland), @d-edison, Dean Attali (@daattali), DanielDavid521 (@Danieldavid521), David Granjon (@DivadNojnarg), Eduardo Vásquez (@edovtp), Emil Hvitfeldt (@EmilHvitfeldt), Emilio (@emilopezcano), Emily Riederer (@emilyriederer), Eric Simms (@esimms999), Federico Marini (@federicomarini), Frederik Kok Hansen (@fkoh111), Frans van Dunné (@FvD), Giorgio Comai (@giocomai), Hedley (@heds1), Henning (@henningsway), Hlynur (@hlynurhallgrims), @hsm207, @jacobxk, James Pooley (@jamespooley), Joe Cheng (@jcheng5), Julien Colomb (@jcolomb), Juan C. Rodriguez (@jcrodriguez1989), Jennifer (Jenny) Bryan (@jennybc), Jim Hester (@jimhester), Joachim Gassen (@joachim-gassen), Jon Calder (@jonmcalder), Jonathan Carroll (@jonocarroll), Julian Stanley (@julianstanley), @jyuu, @kaanpekel, Karandeep Singh (@kdpsingh), Robert Kirk DeLisle (@KirkDCO), Elaine (@loomalaine), Malcolm Barrett (@malcolmbarrett), Marly Gotti (@marlycormar), Matthew Wilson (@MattW-Geospatial), Matthew T. Warkentin (@mattwarkentin), Mauro Lepore (@maurolepore), Maximilian Rohde (@maxdrohde), Matthew Berginski (@mbergins), Michael Dewar (@michael-dewar), Mine Cetinkaya-Rundel (@mine-cetinkaya-rundel), Maria Paula Caldas (@mpaulacaldas), nthobservation (@nthobservation), Pietro Monticone (@pitmonticone), psychometrician (@psychometrician), Ram Thapa (@raamthapa), Janko Thyson (@rappster), Rebecca Janis (@rbjanis), Tom Palmer (@remlapmot), Russ Hyde (@russHyde), Barret Schloerke (@schloerke), Scott (@scottyd22), Matthew Sedaghatfar (@sedaghatfar), Shixiang Wang (@ShixiangWang), Praer (Suthira Owlarn) (@sowla), Sébastien Rochette (@statnmap), @stevensbr, André Calero Valdez (@Sumidu), Tanner Stauss (@tmstauss), Tony Fujs (@tonyfujs), Stefan Moog (@trekonom), Jeff Allen (@trestletech), Trey Gilliland (@treygilliland), Albrecht (@Tungurahua), Valeri Voev (@ValeriVoev), Vickus (@Vickusr), William Doane (@WilDoane), 黄湘云 (@XiangyunHuang), and gXcloud (@xwydq).

How This Book Was Built

This book was written in RStudio (*http://www.rstudio.com/ide*) using bookdown (*http://bookdown.org*). The book's website (*http://mastering-shiny.org*) is hosted with netlify (*http://netlify.com*) and is automatically updated after every commit by GitHub Actions (*https://github.com/features/actions*). The complete source is available from GitHub (*https://github.com/hadley/mastering-shiny*).

This version of the book was built with R version 4.0.3 (2020-10-10) and the following packages:

Package	Version	Source
gapminder	0.3.0	standard (@0.3.0)
ggforce	0.3.2	standard (@0.3.2)
gh	1.2.0	standard (@1.2.0)
globals	0.14.0	standard (@0.14.0)
openintro	2.0.0	standard (@2.0.0)
profvis	0.3.7.9000	GitHub (rstudio/profvis@ca1b272)
RSQLite	2.2.3	standard (@2.2.3)
shiny	1.6.0	standard (@1.6.0)
shinycssloaders	1.0.0	standard (@1.0.0)
shinyFeedback	0.3.0	standard (@0.3.0)
shinythemes	1.2.0	standard (@1.2.0)
testthat	3.0.2.9000	GitHub (r-lib/testthat@4793514)
thematic	0.1.1	GitHub (rstudio/thematic@d78d24a)
tidyverse	1.3.0	standard (@1.3.0)
vroom	1.3.2	standard (@1.3.2)
waiter	0.2.0	standard (@0.2.0)
xml2	1.3.2	standard (@1.3.2)
zeallot	0.1.0	standard (@0.1.0)

## PART I

Getting Started

The goal of the next four chapters is to get you writing Shiny apps as quickly as possible. In Chapter 1, I'll start small, but complete, showing you all the major pieces of an app and how they fit together. Then in Chapters 2 and 3 you'll start to get into the details of the two major parts of a Shiny app: the frontend (what the user sees in the browser) and the backend (the code that makes it all work). We'll finish up in Chapter 4 with a case study to help cement the concepts you've learned so far.

## CHAPTER 1

Your First Shiny App

Introduction

In this chapter, we'll create a simple Shiny app. I'll start by showing you the minimum boilerplate needed for a Shiny app, and then you'll learn how to start and stop it. Next you'll learn the two key components of every Shiny app: the *UI* (short for user interface), which defines how your app *looks*, and the *server function*, which defines how your app *works*. Shiny uses *reactive programming* to automatically update outputs when inputs change, so we'll finish off the chapter by learning the third important component of Shiny apps: reactive expressions.

If you haven't already installed Shiny, install it now with:

```
install.packages("shiny")
```

If you've already installed Shiny, use `packageVersion("shiny")` to check that you have version 1.5.0 or greater.

Then load in your current R session:

```
library(shiny)
```

Create App Directory and File

There are several ways to create a Shiny app. The simplest is to create a new directory for your app and put a single file called *app.R* in it. This *app.R* file will be used to tell Shiny both how your app should look and how it should behave.

Try it out by creating a new directory and adding an *app.R* file that looks like this:

```
library(shiny)
ui <- fluidPage(
  "Hello, world!"
```

```
)
server <- function(input, output, session) {
}
shinyApp(ui, server)
```

This is a complete, if trivial, Shiny app! Looking closely at the preceding code, our *app.R* does four things:

1. It calls `library(shiny)` to load the shiny package.

2. It defines the user interface, the HTML webpage that humans interact with. In this case, it's a page containing the words "Hello, world!"

3. It specifies the behavior of our app by defining a `server` function. It's currently empty, so our app doesn't *do* anything, but we'll be back to revisit this shortly.

4. It executes `shinyApp(ui, server)` to construct and start a Shiny application from the UI and server.

 There are two convenient ways to create a new app in RStudio:

- Create a new directory and an *app.R* file containing a basic app in one step by clicking File → New Project, then selecting New Directory and Shiny Web Application.

- If you've already created the *app.R* file, you can quickly add the app boilerplate by typing **shinyapp** and pressing Shift+Tab.

Running and Stopping

There are a few ways you can run this app:

- Click the Run App (Figure 1-1) button in the document toolbar.

- Use a keyboard shortcut: Cmd/Ctrl+Shift+Enter.

- If you're not using RStudio, you can (source())[1] the whole document or call `shiny::runApp()` with the path to the directory containing *app.R*.

Figure 1-1. The Run App button can be found at the top right of the source pane.

---

[1] The extra () on the outside are important. shinyApp() only creates an app when printed, and () forces the printing of the last result in the file being sourced, which is otherwise returned invisibly.

Pick one of these options, and check that you see the same app as in Figure 1-2. Congratulations! You've made your first Shiny app.

Figure 1-2. They very basic Shiny app you'll see when you run the preceding code.

Before you close the app, go back to RStudio and look at the R console. You'll notice that it says something like:

```
#> Listening on http://127.0.0.1:3827
```

This tells you the URL where your app can be found: 127.0.0.1 is a standard address that means "this computer," and 3827 is a randomly assigned port number. You can enter that URL into any compatible[2] web browser to open another copy of your app.

Also notice that R is busy: the R prompt isn't visible, and the console toolbar displays a stop-sign icon. While a Shiny app is running, it "blocks" the R console. This means that you can't run new commands at the R console until the Shiny app stops.

You can stop the app and return access to the console using any one of these options:

- Click the stop-sign icon on the R console toolbar.
- Click on the console, then press Esc (or press Ctrl+C if you're not using RStudio).
- Close the Shiny app window.

---

2 Shiny strives to support all modern browsers (*https://oreil.ly/CNDxd*). Note that Internet Explorer versions prior to IE11 are not compatible when running Shiny directly from your R session. However, Shiny apps deployed on Shiny Server or ShinyApps.io can work with IE10 (earlier versions of IE are no longer supported).

The basic workflow of Shiny app development is to write some code, start the app, play with the app, write some more code, and repeat. If you're using RStudio, you don't even need to stop and restart the app to see your changes—you can either press the "Reload app" button in the toolbox or use the Cmd/Ctrl+Shift+Enter keyboard shortcut. I'll cover other workflow patterns in Chapter 5.

Adding UI Controls

Next, we'll add some inputs and outputs to our UI so it's not *quite* so minimal. We're going to make a very simple app that shows you all the built-in data frames included in the datasets package.

Replace your ui with this code:

```
ui <- fluidPage(
  selectInput("dataset", label = "Dataset", choices = ls("package:datasets")),
  verbatimTextOutput("summary"),
  tableOutput("table")
)
```

This example uses four new functions:

fluidPage()

> A *layout function* that sets up the basic visual structure of the page. You'll learn more about them in "Single-Page Layouts" on page 89.

selectInput()

> An *input control* that lets the user interact with the app by providing a value. In this case, it's a select box with the label "Dataset" and lets you choose one of the built-in datasets that come with R. You'll learn more about inputs in "Inputs" on page 15.

verbatimTextOutput() *and* tableOutput()

> *Output controls* that tell Shiny *where* to put rendered output (we'll get into the *how* in a moment). verbatimTextOutput() displays code, and tableOutput() displays tables. You'll learn more about outputs in "Outputs" on page 22.

Layout functions, inputs, and outputs have different uses, but they are fundamentally the same under the covers: they're all just fancy ways to generate HTML, and if you call any of them outside of a Shiny app, you'll see HTML printed out at the console. Don't be afraid to poke around to see how these various layouts and controls work under the hood.

Go ahead and run the app again. You'll now see what appears in Figure 1-3, a page containing a select box. We only see the input, not the two outputs, because we haven't yet told Shiny how the input and outputs are related.

Figure 1-3. The datasets app with UI.

Adding Behavior

Next, we'll bring the outputs to life by defining them in the server function.

Shiny uses reactive programming to make apps interactive. You'll learn more about reactive programming in Chapter 3, but for now, just be aware that it involves telling Shiny *how* to perform a computation, not ordering Shiny to actually go *do it*. It's like the difference between giving someone a recipe and demanding that they go make you a sandwich.

We'll tell Shiny how to fill in the `summary` and `table` outputs in the sample app by providing the "recipes" for those outputs. Replace your empty `server` function with this:

```
server <- function(input, output, session) {
  output$summary <- renderPrint({
    dataset <- get(input$dataset, "package:datasets")
    summary(dataset)
  })

  output$table <- renderTable({
    dataset <- get(input$dataset, "package:datasets")
    dataset
  })
}
```

The left-hand side of the assignment operator (`<-`), `output$ID`, indicates that you're providing the recipe for the Shiny output with that ID. The right-hand side of the assignment uses a specific *render function* to wrap some code that you provide. Each `render{Type}` function is designed to produce a particular type of output (e.g., text, tables, and plots) and is often paired with a `{type}Output` function. For example, in this app, `renderPrint()` is paired with `verbatimTextOutput()` to display a statistical summary with fixed-width (verbatim) text, and `renderTable()` is paired with `tableOutput()` to show the input data in a table.

Run the app again and play around, watching what happens to the output when you change an input. Figure 1-4 shows what you should see when you open the app.

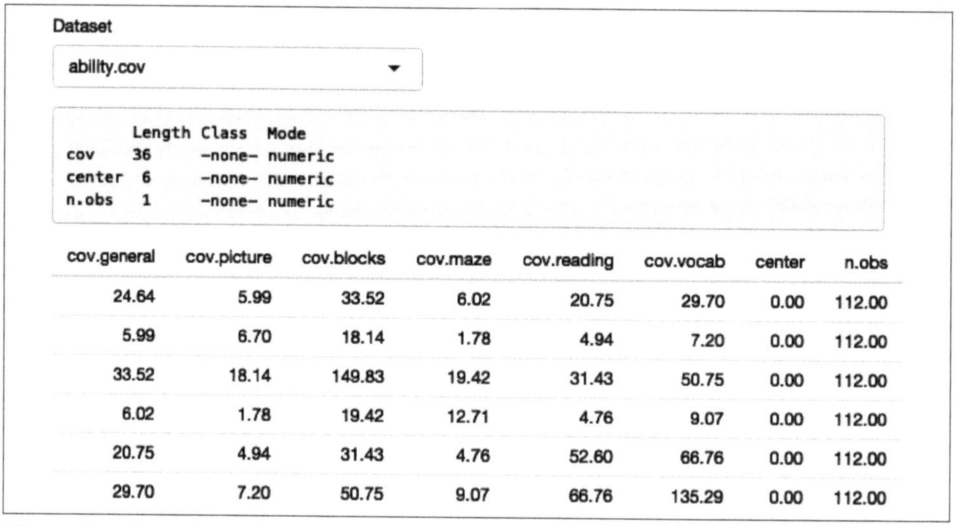

Dataset

ability.cov ▼

```
        Length Class  Mode
cov     36     -none- numeric
center  6      -none- numeric
n.obs   1      -none- numeric
```

cov.general	cov.picture	cov.blocks	cov.maze	cov.reading	cov.vocab	center	n.obs
24.64	5.99	33.52	6.02	20.75	29.70	0.00	112.00
5.99	6.70	18.14	1.78	4.94	7.20	0.00	112.00
33.52	18.14	149.83	19.42	31.43	50.75	0.00	112.00
6.02	1.78	19.42	12.71	4.76	9.07	0.00	112.00
20.75	4.94	31.43	4.76	52.60	66.76	0.00	112.00
29.70	7.20	50.75	9.07	66.76	135.29	0.00	112.00

Figure 1-4. Now that we've provided a server function that connects outputs and inputs, we have a fully functional app.

Notice that the summary and table update whenever you change the input dataset. This dependency is created implicitly because we've referred to `input$dataset` within the output functions. `input$dataset` is populated with the current value of the UI component with ID `dataset` and will cause the outputs to automatically update whenever that value changes. This is the essence of *reactivity*: outputs automatically *react* (recalculate) when their inputs change.

Reducing Duplication with Reactive Expressions

Even in this simple example, we have some code that's duplicated: the following line is present in both outputs:

```
dataset <- get(input$dataset, "package:datasets")
```

In every kind of programming, it's poor practice to have duplicated code; it can be computationally wasteful, and more importantly, it increases the difficulty of maintaining or debugging the code. It's not that important here, but I wanted to illustrate the basic idea in a very simple context.

In traditional R scripting, we use two techniques to deal with duplicated code: either we capture the value using a variable or we capture the computation with a function. Unfortunately, neither of these approaches works here, for reasons you'll learn about in "Why Do We Need Reactive Programming?" on page 202, and we need a new mechanism: *reactive expressions*.

You create a reactive expression by wrapping a block of code in reactive({...}) and assigning it to a variable, and you use a reactive expression by calling it like a function. But while it looks like you're calling a function, a reactive expression has an important difference: it only runs the first time it is called, and then it caches its result until it needs to be updated.

We can update our server() to use reactive expressions, as shown in the following code. The app behaves identically but works a little more efficiently because it only needs to retrieve the dataset once, not twice:

```
server <- function(input, output, session) {
  # Create a reactive expression
  dataset <- reactive({
    get(input$dataset, "package:datasets")
  })

  output$summary <- renderPrint({
    # Use a reactive expression by calling it like a function
    summary(dataset())
  })

  output$table <- renderTable({
    dataset()
  })
}
```

We'll come back to reactive programming multiple times, but even armed with a cursory knowledge of inputs, outputs, and reactive expressions, it's possible to build quite useful Shiny apps!

Summary

In this chapter you've created a simple app—it's not very exciting or useful, but you can see how easy it is to construct a web app using your existing R knowledge. In the next two chapters, you'll learn more about user interfaces and reactive programming, the two basic building blocks of Shiny. Now is a great time to grab a copy of the Shiny cheat sheet (*https://oreil.ly/vJbBe*). This is a great resource to help jog your memory of the main components of a Shiny app.

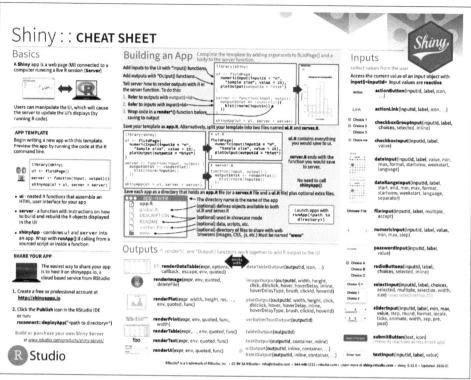

Figure 1-5. *The Shiny cheat sheet (https://www.rstudio.com/resources/cheatsheets).*

Exercises

1. Create an app that greets the user by name. You don't know all the functions you need to do this yet, so I've included some lines of code here. Think about which lines you'll use and then copy and paste them into the right place in a Shiny app:

```
tableOutput("mortgage")
output$greeting <- renderText({
  paste0("Hello ", input$name)
})
numericInput("age", "How old are you?", value = NA)
textInput("name", "What's your name?")
textOutput("greeting")
output$histogram <- renderPlot({
  hist(rnorm(1000))
}, res = 96)
```

2. Suppose your friend wants to design an app that allows the user to set a number (x) between 1 and 50 and displays the result of multiplying this number by 5. This is their first attempt:

```
library(shiny)

ui <- fluidPage(
  sliderInput("x", label = "If x is", min = 1, max = 50, value = 30),
  "then x times 5 is",
  textOutput("product")
)

server <- function(input, output, session) {
  output$product <- renderText({
    x * 5
  })
}

shinyApp(ui, server)
```

But unfortunately it has an error:

then x times 5 is
Error: object 'x' not found

Can you help them find and correct the error?

3. Extend the app from the previous exercise to allow the user to set the value of the multiplier, y, so that the app yields the value of x * y. The final result should look like this:.

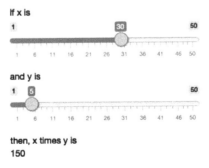

4. Take the following app, which adds some additional functionality to the last app described in the last exercise. What's new? How could you reduce the amount of duplicated code in the app by using a reactive expression?

```r
library(shiny)

ui <- fluidPage(
  sliderInput("x", "If x is", min = 1, max = 50, value = 30),
  sliderInput("y", "and y is", min = 1, max = 50, value = 5),
  "then, (x * y) is", textOutput("product"),
  "and, (x * y) + 5 is", textOutput("product_plus5"),
  "and (x * y) + 10 is", textOutput("product_plus10")
)

server <- function(input, output, session) {
  output$product <- renderText({
    product <- input$x * input$y
    product
  })
  output$product_plus5 <- renderText({
    product <- input$x * input$y
    product + 5
  })
  output$product_plus10 <- renderText({
    product <- input$x * input$y
    product + 10
  })
}

shinyApp(ui, server)
```

5. The following app is very similar to one you've seen earlier in the chapter: you select a dataset from a package (this time we're using the ggplot2 package), and the app prints out a summary and plot of the data. It also follows good practice and makes use of reactive expressions to avoid redundancy of code. However, there are three bugs in the following code. Can you find and fix them?

```r
library(shiny)
library(ggplot2)

datasets <- c("economics", "faithfuld", "seals")
ui <- fluidPage(
  selectInput("dataset", "Dataset", choices = datasets),
  verbatimTextOutput("summary"),
  tableOutput("plot")
)

server <- function(input, output, session) {
  dataset <- reactive({
    get(input$dataset, "package:ggplot2")
```

```
  })
  output$summary <- renderPrint({
    summary(dataset())
  })
  output$plot <- renderPlot({
    plot(dataset)
  }, res = 96)
}

shinyApp(ui, server)
```

## CHAPTER 2

Basic UI

Introduction

Now that you have a basic app under your belt, we can start to explore the details that make Shiny tick. As you saw in the previous chapter, Shiny encourages separation of the code that generates your user interface (the frontend) from the code that drives your app's behavior (the backend).

In this chapter, we'll focus on the frontend and give you a whirlwind tour of the HTML inputs and outputs provided by Shiny. This gives you the ability to capture many types of data and display many types of R output. You don't yet have many ways to stitch the inputs and outputs together, but we'll come back to that in Chapter 6.

Here I'll mostly stick to the inputs and outputs built into Shiny itself. However, there is a rich and vibrant community of extension packages, like shinyWidgets (*https:// oreil.ly/7WHmU*), colourpicker (*https://oreil.ly/sFdk6*), and sortable (*https://rstu dio.github.io/sortable*). Nan Xiao (*https://nanx.me*) maintains an active and comprehensive list of other Shiny packages (*https://oreil.ly/t2TQ9*).

As usual, we'll begin by loading the Shiny package:

```
library(shiny)
```

Inputs

As we saw in the previous chapter, you use functions like `sliderInput()`, `selectIn put()`, `textInput()`, and `numericInput()` to insert input controls into your UI specification. Now we'll discuss the common structure that underlies all input functions and give a quick overview of the inputs built into Shiny.

Common Structure

All input functions have the same first argument: `inputId`. This is the identifier used to connect the frontend with the backend: if your UI has an input with ID `"name"`, the server function will access it with `input$name`.

The `inputId` has two constraints:

- It must be a simple string that contains only letters, numbers, and underscores (no spaces, dashes, periods, or other special characters allowed!). Name it like you would name a variable in R.

- It must be unique. If it's not unique, you'll have no way to refer to this control in your server function!

Most input functions have a second parameter called `label`. This is used to create a human-readable label for the control. Shiny doesn't place any restrictions on this string, but you'll need to carefully think about it to make sure that your app is usable by humans! The third parameter is typically `value`, which, where possible, lets you set the default value. The remaining parameters are unique to the control.

When creating an input, I recommend supplying the `inputId` and `label` arguments by position and all other arguments by name:

```
sliderInput("min", "Limit (minimum)", value = 50, min = 0, max = 100)
```

The following sections describe the inputs built into Shiny, loosely grouped according to the type of control they create. The goal is to give you a rapid overview of your options, not to exhaustively describe all the arguments. I'll show the most important parameters for each control here, but you'll need to read the documentation to get the full details.

Free Text

Collect small amounts of text with `textInput()`, passwords with `passwordInput()`,[1] and paragraphs of text with `textAreaInput()`:

```
ui <- fluidPage(
  textInput("name", "What's your name?"),
  passwordInput("password", "What's your password?"),
  textAreaInput("story", "Tell me about yourself", rows = 3)
)
```

---

1 All `passwordInput()` does is hide what the user is typing so that someone looking over their shoulder can't read it. It's up to you to make sure that any passwords are not accidentally exposed, so we don't recommend using passwords unless you have had some training in secure programming.

What's your name?

What's your password?

Tell me about yourself

If you want to ensure that the text has certain properties, you can use `validate()`, which we'll come back to in Chapter 8.

Numeric Inputs

To collect numeric values, create a constrained text box with `numericInput()` or a slider with `sliderInput()`. If you supply a length-2 numeric vector for the default value of `sliderInput()`, you get a "range" slider with two ends:

```
ui <- fluidPage(
  numericInput("num", "Number one", value = 0, min = 0, max = 100),
  sliderInput("num2", "Number two", value = 50, min = 0, max = 100),
  sliderInput("rng", "Range", value = c(10, 20), min = 0, max = 100)
)
```

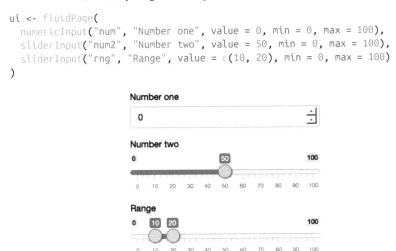

Generally, I recommend only using sliders for small ranges or cases where the precise value is not so important. Attempting to precisely select a number on a small slider is an exercise in frustration!

Sliders are extremely customizable, and there are many ways to tweak their appearance. See `?sliderInput` and "Using sliders" (*https://oreil.ly/BLsCJ*) for more details.

Dates

Collect a single day with `dateInput()` or a range of two days with `dateRangeInput()`. These functions provide a convenient calendar picker, and additional arguments like `datesdisabled` and `daysofweekdisabled` allow you to restrict the set of valid inputs:

```
ui <- fluidPage(
  dateInput("dob", "When were you born?"),
  dateRangeInput("holiday", "When do you want to go on vacation next?")
)
```

Date format, language, and the day on which the week starts defaults to US standards. If you are creating an app with an international audience, set `format`, `language`, and `weekstart` so that the dates are natural to your users.

Limited Choices

There are two different approaches to allow the user to choose from a prespecified set of options: `selectInput()` and `radioButtons()`:

```
animals <- c("dog", "cat", "mouse", "bird", "other", "I hate animals")

ui <- fluidPage(
  selectInput("state", "What's your favourite state?", state.name),
  radioButtons("animal", "What's your favourite animal?", animals)
)
```

Radio buttons have two nice features: they show all possible options, making them suitable for short lists, and via the `choiceNames`/`choiceValues` arguments, they can display options other than plain text. `choiceNames` determines what is shown to the user; `choiceValues` determines what is returned in your server function:

```
ui <- fluidPage(
  radioButtons("rb", "Choose one:",
    choiceNames = list(
      icon("angry"),
      icon("smile"),
      icon("sad-tear")
    ),
    choiceValues = list("angry", "happy", "sad")
  )
)
```

Choose one:

Drop-downs created with `selectInput()` take up the same amount of space, regardless of the number of options, making them more suitable for longer options. You can also set `multiple = TRUE` to allow the user to select multiple elements:

```
ui <- fluidPage(
  selectInput(
    "state", "What's your favourite state?", state.name,
    multiple = TRUE
  )
)
```

What's your favourite state?

| Texas Cal |
| California |

If you have a very large set of possible options, you may want to use "server-side" `selectInput()` so that the complete set of possible options is not embedded in the UI (which can make it slow to load) but instead sent as needed by the server. You can learn more about this advanced topic in the Shiny documentation (*https://oreil.ly/FgvCM*).

There's no way to select multiple values with radio buttons, but there's an alternative that's conceptually similar: `checkboxGroupInput()`:

```
ui <- fluidPage(
  checkboxGroupInput("animal", "What animals do you like?", animals)
)
```

What animals do you like?

☐ dog

☐ cat

☐ mouse

☐ bird

☐ other

☐ I hate animals

If you want a single checkbox for a single yes/no question, use checkboxInput():

```
ui <- fluidPage(
  checkboxInput("cleanup", "Clean up?", value = TRUE),
  checkboxInput("shutdown", "Shutdown?")
)
```

☑ Clean up?

☐ Shutdown?

File Uploads

Allow the user to upload a file with fileInput():

```
ui <- fluidPage(
  fileInput("upload", NULL)
)
```

| Browse... | No file selected |

fileInput() requires special handling on the server side and is discussed in detail in Chapter 9.

Action Buttons

Let the user perform an action with actionButton() or actionLink():

```
ui <- fluidPage(
  actionButton("click", "Click me!"),
  actionButton("drink", "Drink me!", icon = icon("cocktail"))
)
```

Click me!   🍸 Drink me!

Action links and buttons are most naturally paired with `observeEvent()` or even `tReactive()` in your server function. You haven't learned about these important functions yet, but we'll come back to them in "Controlling Timing of Evaluation" on page 44.

You can customize the appearance using the `class` argument by using one of `"btn-primary"`, `"btn-success"`, `"btn-info"`, `"btn-warning"`, or `"btn-danger"`. You can also change the size with `"btn-lg"`, `"btn-sm"`, or `"btn-xs"`. Finally, you can make buttons span the entire width of the element they are embedded within using `"btn-block"`:

```r
ui <- fluidPage(
  fluidRow(
    actionButton("click", "Click me!", class = "btn-danger"),
    actionButton("drink", "Drink me!", class = "btn-lg btn-success")
  ),
  fluidRow(
    actionButton("eat", "Eat me!", class = "btn-block")
  )
)
```

The `class` argument works by setting the `class` attribute of the underlying HTML, which affects how the element is styled. To see other options, you can read the documentation for Bootstrap (*https://oreil.ly/6VHyv*), the CSS design system used by Shiny.

Exercises

1. When space is at a premium, it's useful to label text boxes using a placeholder that appears *inside* the text entry area. How do you call `textInput()` to generate the following UI?

2. Carefully read the documentation for `sliderInput()` to figure out how to create a date slider, as shown here:

3. Create a slider input to select values between 0 and 100 where the interval between each selectable value on the slider is 5. Then, add animation to the input widget so when the user presses play, the input widget scrolls through the range automatically.

4. If you have a moderately long list in a `selectInput()`, it's useful to create sub-headings that break the list up into pieces. Read the documentation to figure out how. (Hint: The underlying HTML is called `<optgroup>`.)

Outputs

Outputs in the UI create placeholders that are later filled by the server function. Like inputs, outputs take a unique ID as their first argument:[2] if your UI specification creates an output with ID `"plot"`, you'll access it in the server function with `output$plot`.

Each `output` function on the frontend is coupled with a `render` function in the backend. There are three main types of output, corresponding to the three things you usually include in a report: text, tables, and plots. The following sections show you the basics of the output functions on the frontend, along with the corresponding `render` functions in the backend.

Text

Output regular text with `textOutput()` and fixed width text (e.g., console output) with `verbatimTextOutput()`:

```
ui <- fluidPage(
  textOutput("text"),
  verbatimTextOutput("code")
)
server <- function(input, output, session) {
  output$text <- renderText({
    "Hello friend!"
  })
  output$code <- renderPrint({
    summary(1:10)
  })
}
```

---

2 Note that the name of that argument is different for inputs (`inputId`) and outputs (`outputId`). I don't use the name of the first argument because it's so important, and I expect you to remember what it does without an additional hint.

```
Hello friend!

 Min. 1st Qu.  Median    Mean 3rd Qu.    Max.
 1.00    3.25    5.50    5.50    7.75   10.00
```

Note that the {} are only required in render functions if you need to run multiple lines of code. As you'll learn shortly, you should do as little computation in your render functions as possible, which means you can often omit them. Here's what the preceding server function would look like if written more compactly:

```
server <- function(input, output, session) {
  output$text <- renderText("Hello friend!")
  output$code <- renderPrint(summary(1:10))
}
```

Note that there are two render functions, which behave slightly differently:

renderText()

This combines the result into a single string and is usually paired with textOutput().

renderPrint()

This *prints* the result, as if you were in an R console, and is usually paired with verbatimTextOutput().

We can see the difference with a toy app:

```
ui <- fluidPage(
  textOutput("text"),
  verbatimTextOutput("print")
)
server <- function(input, output, session) {
  output$text <- renderText("hello!")
  output$print <- renderPrint("hello!")
}
```

```
hello!

[1] "hello!"
```

This is equivalent to the difference between cat() and print() in base R.

Tables

There are two options for displaying data frames in tables:

`tableOutput()` *and* `renderTable()`
These render a static table of data, showing all the data at once.

`dataTableOutput()` *and* `renderDataTable()`
These render a dynamic table, showing a fixed number of rows along with controls to change which rows are visible.

`tableOutput()` is most useful for small, fixed summaries (e.g., model coefficients); `dataTableOutput()` is most appropriate if you want to expose a complete data frame to the user. The following code shows a very simple example of `tableOutput()` and `dataTableOutput()`:

```
ui <- fluidPage(
  tableOutput("static"),
  dataTableOutput("dynamic")
)
server <- function(input, output, session) {
  output$static <- renderTable(head(mtcars))
  output$dynamic <- renderDataTable(mtcars, options = list(pageLength = 5))
}
```

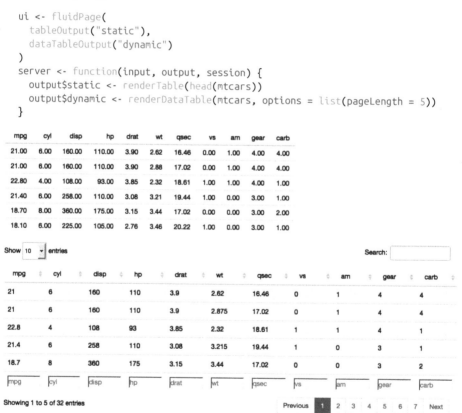

If you want greater control over the output of `dataTableOutput()`, I highly recommend the reactable (*https://glin.github.io/reactable*) package by Greg Lin.

Plots

You can display any type of R graphic (e.g., base or ggplot2) with `plotOutput()` and `renderPlot()`:

```
ui <- fluidPage(
  plotOutput("plot", width = "400px")
)
server <- function(input, output, session) {
  output$plot <- renderPlot(plot(1:5), res = 96)
}
```

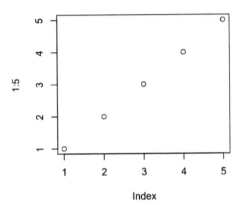

By default, `plotOutput()` will take up the full width of its container (more on that shortly) and will be 400 pixels high. You can override these defaults with the `height` and `width` arguments. We recommend always setting `res = 96` as that will make your Shiny plots match what you see in RStudio as closely as possible.

Plots are special because they are outputs that can also act as inputs. `plotOutput()` has a number of arguments like `click`, `dblclick`, and `hover`. If you pass these a string, like `click = "plot_click"`, they'll create a reactive input (`input$plot_click`) that you can use to handle user interaction on the plot (e.g., clicking on the plot). We'll come back to interactive plots in Shiny in Chapter 7.

Downloads

You can let the user download a file with `downloadButton()` or `downloadLink()`. These require new techniques in the server function, so we'll come back to that in Chapter 9.

Exercises

1. Which of `textOutput()` and `verbatimTextOutput()` should each of the following render functions be paired with?

   a. `renderPrint(summary(mtcars))`

   b. `renderText("Good morning!")`

   c. `renderPrint(t.test(1:5, 2:6))`

   d. `renderText(str(lm(mpg ~ wt, data = mtcars)))`

2. Re-create the Shiny app from "Plots" on page 25, this time setting height to 300px and width to 700px. Set the plot "alt" text so that a visually impaired user can tell that it's a scatterplot of five random numbers.

3. Update the options in the call to `renderDataTable()` so that the data is displayed but all other controls are suppressed (i.e., remove the search, ordering, and filtering commands). You'll need to read `?renderDataTable` and review the options in the underlying JavaScript library (*https://datatables.net/reference/option*):

   ```
   ui <- fluidPage(
     dataTableOutput("table")
   )
   server <- function(input, output, session) {
     output$table <- renderDataTable(mtcars, options = list(pageLength = 5))
   }
   ```

4. Alternatively, read up on reactable (*https://glin.github.io/reactable*) and convert the preceding app to use it instead.

Summary

This chapter has introduced you to the major input and output functions that make up the frontend of a Shiny app. This was a big infodump, so don't expect to remember everything after a single read. Instead, come back to this chapter when you're looking for a specific component: you can quickly scan the figures and then find the code you need.

In the next chapter, we'll move on to the backend of a Shiny app: the R code that makes your user interface come to life.

## CHAPTER 3

Basic Reactivity

Introduction

In Shiny, you express your server logic using reactive programming. Reactive programming is an elegant and powerful programming paradigm, but it can be disorienting at first because it's a very different paradigm to writing a script. The key idea of reactive programming is to specify a graph of dependencies so that when an input changes, all related outputs are automatically updated. This makes the flow of an app considerably simpler, but it takes a while to get your head around how it all fits together.

This chapter will provide a gentle introduction to reactive programming, teaching you the basics of the most common reactive constructs you'll use in Shiny apps. We'll start with a survey of the server function, discussing in more detail how the input and output arguments work. Next we'll review the simplest form of reactivity (where inputs are directly connected to outputs) and then discuss how reactive expressions allow you to eliminate duplicated work. We'll finish by reviewing some common roadblocks encountered by newer Shiny users.

The Server Function

As you've seen, the guts of every Shiny app look like this:

```r
library(shiny)

ui <- fluidPage(
  # frontend interface
)

server <- function(input, output, session) {
  # backend logic
```

```
}
shinyApp(ui, server)
```

The previous chapter covered the basics of the frontend, the ui object that contains the HTML presented to every user of your app. The ui is simple because every user gets the same HTML. The server is more complicated because every user needs to get an independent version of the app; when user A moves a slider, user B shouldn't see their outputs change.

To achieve this independence, Shiny invokes your server() function each time a new session starts.[1] Just like any other R function, when the server function is called, it creates a new local environment that is independent of every other invocation of the function. This allows each session to have a unique state and isolate the variables created *inside* the function. This is why almost all of the reactive programming you'll do in Shiny will be inside the server function.[2]

Server functions take three parameters: input, output, and session. Because you never call the server function yourself, you'll never create these objects yourself. Instead, they're created by Shiny when the session begins, connecting back to a specific session. For the moment, we'll focus on the input and output arguments and leave session for later chapters.

Input

The input argument is a list-like object that contains all the input data sent from the browser, named according to the input ID. For example, if your UI contains a numeric input control with an input ID of count, like so:

```
ui <- fluidPage(
  numericInput("count", label = "Number of values", value = 100)
)
```

then you can access the value of that input with input$count. It will initially contain the value 100, and it will be automatically updated as the user changes the value in the browser.

---

1 Each connection to a Shiny app starts a new session, whether it's connections from different people or multiple tabs from the same person.

2 The primary exception is where there's some work that can be shared across multiple users. For example, all users might be looking at the same large CSV file, so you might as well load it once and share it between users. We'll come back to that idea in "Schedule Data Munging" on page 338.

Unlike a typical list, `input` objects are read-only. If you attempt to modify an input inside the server function, you'll get an error:

```
server <- function(input, output, session) {
  input$count <- 10
}

shinyApp(ui, server)
#> Error: Can't modify read-only reactive value 'count'
```

This error occurs because `input` reflects what's happening in the browser, and the browser is Shiny's "single source of truth." If you could modify the value in R, you could introduce inconsistencies, where the input slider said one thing in the browser and `input$count` said something different in R. That would make programming challenging! Later, in Chapter 8, you'll learn how to use functions like `updateNumericInput()` to modify the value in the browser, and then `input$count` will update accordingly.

One more important thing about `input`: it's selective about who is allowed to read it. To read from an `input`, you must be in a *reactive context* created by a function like `renderText()` or `reactive()`. We'll come back to that idea very shortly, but it's an important constraint that allows outputs to automatically update when an input changes. This code illustrates the error you'll see if you make this mistake:

```
server <- function(input, output, session) {
  message("The value of input$count is ", input$count)
}

shinyApp(ui, server)
#> Error: Can't access reactive value 'count' outside of reactive consumer.
#> i Do you need to wrap inside reactive() or observer()?
```

Output

`output` is very similar to `input`: it's also a list-like object named according to the output ID. The main difference is that you use it for sending output instead of receiving input. You always use the `output` object in concert with a `render` function, as in the following simple example:

```
ui <- fluidPage(
  textOutput("greeting")
)

server <- function(input, output, session) {
  output$greeting <- renderText("Hello human!")
}
```

(Note that the ID is quoted in the UI but not in the server.)

The render function does two things:

- It sets up a special reactive context that automatically tracks what inputs the output uses.
- It converts the output of your R code into HTML suitable for display on a web page.

Like the input, the output is picky about how you use it. You'll get an error if:

- You forget the render function:

```
server <- function(input, output, session) {
  output$greeting <- "Hello human"
}
shinyApp(ui, server)
#> Error: Unexpected character object for output$greeting
#> i Did you forget to use a render function?
```

- You attempt to read from an output:

```
server <- function(input, output, session) {
  message("The greeting is ", output$greeting)
}
shinyApp(ui, server)
#> Error: Reading from shinyoutput object is not allowed.
```

Reactive Programming

An app is going to be pretty boring if it only has inputs or only has outputs. The real magic of Shiny happens when you have an app with both. Let's look at a simple example:

```
ui <- fluidPage(
  textInput("name", "What's your name?"),
  textOutput("greeting")
)

server <- function(input, output, session) {
  output$greeting <- renderText({
    paste0("Hello ", input$name, "!")
  })
}
```

It's hard to show how this works in a book, but I do my best in Figure 3-1. If you run the app, and type in the name box, you'll see that the greeting updates automatically as you type.[3]

Figure 3-1. Reactivity means that outputs automatically update as inputs change, as in this app where I type: *J, o, e*. See live at https://hadley.shinyapps.io/ms-connection.

This is the big idea in Shiny: you don't need to tell an output when to update, because Shiny automatically figures it out for you. How does it work? What exactly is going on in the body of the function? Let's think about the code inside the server function more precisely:

```
output$greeting <- renderText({
  paste0("Hello ", input$name, "!")
})
```

It's easy to read this as "paste together 'hello' and the user's name, then send the rendered text to output$greeting." But this mental model is wrong in a subtle, but important, way. Think about it: with this model, you only issue the instruction once. But Shiny performs the action every time we update input$name, so there must be something more going on.

The app works because the code doesn't *tell* Shiny to create the string and send it to the browser but instead informs Shiny *how it could* create the string if it needs to. It's up to Shiny when (and even if!) the code should be run. It might be run as soon as the app launches, or it might be quite a bit later; it might be run many times, or it might never be run! This isn't to imply that Shiny is capricious, only that it's Shiny's responsibility to decide when code is executed, not yours. Think of your app as providing Shiny with recipes, not giving it commands.

Imperative Versus Declarative Programming

This difference between commands and recipes is one of the key differences between two important styles of programming:

---

3 If you're running the live app, notice that you have to type fairly slowly for the output to update one letter at a time. That's because Shiny uses a technique called *debouncing*, which means that it waits for a few ms before sending an update. That considerably reduces the amount of work that Shiny needs to do, without appreciably reducing the response time of the app.

Imperative programming

Issue a specific command and it's carried out immediately. This is the style of programming you're used to in your analysis scripts: you command R to load your data, transform it, visualize it, and save the results to disk.

Declarative programming

Express higher-level goals or describe important constraints, and rely on someone else to decide how and/or when to translate that into action. This is the style of programming you use in Shiny.

With imperative code, you say, "Make me a sandwich."[4] With declarative code, you say, "Ensure there is a sandwich in the refrigerator whenever I look inside of it." Imperative code is assertive; declarative code is passive-aggressive.

Most of the time, declarative programming is tremendously freeing: you describe your overall goals, and the software figures out how to achieve them without further intervention. The downside is the occasional time when you know exactly what you want, but you can't figure out how to frame it in a way that the declarative system understands.[5] The goal of this book is to help you develop your understanding of the underlying theory so that happens as infrequently as possible.

Laziness

One of the strengths of declarative programming in Shiny is that it allows apps to be extremely lazy. A Shiny app will only ever do the minimal amount of work needed to update the output controls that you can currently see.[6] This laziness, however, comes with an important downside that you should be aware of. Can you spot what's wrong with the following server function?

```
server <- function(input, output, session) {
  output$greeting <- renderText({
    paste0("Hello ", input$name, "!")
  })
}
```

If you look closely, you might notice that I've written `greting` instead of `greeting`. This won't generate an error in Shiny, but it won't do what you want. The `greting` output doesn't exist, so the code inside `renderText()` will never be run.

---

4 Read this xkcd comic (*https://xkcd.com/149*) for reference.

5 If you've ever struggled to get a ggplot2 legend to look exactly the way you want, you've encountered this problem!

6 Yes, Shiny doesn't update the output if you can't see it in your browser! Shiny is so lazy that it doesn't do the work unless you can actually see the results.

If you're working on a Shiny app and you just can't figure out why your code never gets run, double-check that your UI and server functions are using the same identifiers.

The Reactive Graph

Shiny's laziness has another important property. In most R code, you can understand the order of execution by reading the code from top to bottom. That doesn't work in Shiny, because code is only run when needed. To understand the order of execution, you need to instead look at the *reactive graph*, which describes how inputs and outputs are connected. The reactive graph for the preceding app is very simple and shown in Figure 3-2.

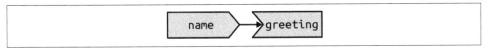

Figure 3-2. The reactive graph shows how the inputs and outputs are connected.

The reactive graph contains one symbol for every input and output, and we connect an input to an output whenever the output accesses the input. This graph tells you that greeting will need to be recomputed whenever name is changed. To describe this relationship, we'll often say that greeting has a *reactive dependency* on name.

Note the graphical conventions we used for the inputs and outputs: the name input naturally fits into the greeting output. We could draw them closely packed together, as in Figure 3-3, to emphasize the way that they fit together; we won't normally do that because it only works for the simplest of apps.

Figure 3-3. The shapes used by the components of the reactive graph evoke the ways in which they connect.

The reactive graph is a powerful tool for understanding how your app works. As your app gets more complicated, it's often useful to make a quick high-level sketch of the reactive graph to remind you how all the pieces fit together. Throughout this book we'll show you the reactive graph to help understand how the examples work, and later on, in Chapter 14, you'll learn how to use reactlog, which will draw the graph for you.

Reactive Expressions

There's one more important component that you'll see in the reactive graph: the reactive expression. We'll come back to the reactive expression in detail very shortly; for now, think of it as a tool that reduces duplication in your reactive code by introducing additional nodes into the reactive graph.

We don't need a reactive expression in our very simple app, but I'll add one anyway so you can see how it affects the reactive graph, Figure 3-4:

```
server <- function(input, output, session) {
  string <- reactive(paste0("Hello ", input$name, "!"))
  output$greeting <- renderText(string())
}
```

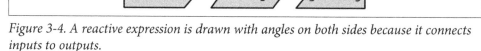

Figure 3-4. A reactive expression is drawn with angles on both sides because it connects inputs to outputs.

Reactive expressions take inputs and produce outputs so they have a shape that combines features of both inputs and outputs. Hopefully, the shapes will help you remember how the components fit together.

Execution Order

It's important to understand that the order in which your code is run is determined solely by the reactive graph. This is different from most R code where the execution order is determined by the order of lines. For example, we could flip the order of the two lines in our simple server function:

```
server <- function(input, output, session) {
  output$greeting <- renderText(string())
  string <- reactive(paste0("Hello ", input$name, "!"))
}
```

You might think that this would yield an error because output$greeting refers to a reactive expression, string, that hasn't been created yet. But remember that Shiny is lazy, so that code is only run when the session starts, after string has been created.

This code yields the same reactive graph as before, so the order in which the code is run is exactly the same. But organizing your code like this is confusing for humans and best avoided. Instead, make sure that reactive expressions and outputs only refer to things defined above, not below.[7] This will make your code easier to understand.

This concept is very important and different from most other R code, so I'll say it again: the order in which reactive code is run is determined only by the reactive graph, not by its layout in the server function.

---

7 The technical term for this ordering is a *topological sort*.

Exercises

1. Given this UI:

```
ui <- fluidPage(
  textInput("name", "What's your name?"),
  textOutput("greeting")
)
```

fix the simple errors found in each of the following three server functions. First try spotting the problem just by reading the code; then run the code to make sure you've fixed it:

```
server1 <- function(input, output, server) {
  input$greeting <- renderText(paste0("Hello ", name))
}

server2 <- function(input, output, server) {
  greeting <- paste0("Hello ", input$name)
  output$greeting <- renderText(greeting)
}

server3 <- function(input, output, server) {
  output$greting <- paste0("Hello", input$name)
}
```

2. Draw the reactive graph for the following server functions:

```
server1 <- function(input, output, session) {
  c <- reactive(input$a + input$b)
  e <- reactive(c() + input$d)
  output$f <- renderText(e())
}
server2 <- function(input, output, session) {
  x <- reactive(input$x1 + input$x2 + input$x3)
  y <- reactive(input$y1 + input$y2)
  output$z <- renderText(x() / y())
}
server3 <- function(input, output, session) {
  d <- reactive(c() ^ input$d)
  a <- reactive(input$a * 10)
  c <- reactive(b() / input$c)
  b <- reactive(a() + input$b)
}
```

3. Why will this code fail?

```
var <- reactive(df[[input$var]])
range <- reactive(range(var(), na.rm = TRUE))
```

Why are range() and var() bad names for reactive?

Reactive Expressions

We've quickly skimmed over reactive expressions a couple of times, so you're hopefully getting a sense of what they might do. Now we'll dive into more of the details and show why they are so important when constructing real apps.

Reactive expressions are important because they give *Shiny* more information so that it can do less recomputation when inputs change, making apps more efficient, and they make it easier for *humans* to understand the app by simplifying the reactive graph. Reactive expressions have a flavor of both inputs and outputs:

- Like inputs, you can use the results of a reactive expression in an output.
- Like outputs, reactive expressions depend on inputs and automatically know when they need updating.

This duality means we need some new vocab: I'll use *producers* to refer to reactive inputs and expressions, and *consumers* to refer to reactive expressions and outputs. Figure 3-5 shows this relationship with a Venn diagram.

Figure 3-5. Inputs and expressions are reactive producers; expressions and outputs are reactive consumers.

We're going to need a more complex app to see the benefits of using reactive expressions. First, we'll set the stage by defining some regular R functions that we'll use to power our app.

The Motivation

Imagine I want to compare two simulated datasets with a plot and a hypothesis test. I've done a little experimentation and come up with the following functions: `freqp oly()` visualizes the two distributions with frequency polygons,[8] and `t_test()` uses a t-test to compare means and summarizes the results with a string:

```
library(ggplot2)

freqpoly <- function(x1, x2, binwidth = 0.1, xlim = c(-3, 3)) {
  df <- data.frame(
```

---

[8] If you haven't heard of a frequency polygon before, it's just a histogram that's drawn with a line instead of bars, which makes it easier to compare multiple datasets on the same plot.

```
    x = c(x1, x2),
    g = c(rep("x1", length(x1)), rep("x2", length(x2)))
  )

  ggplot(df, aes(x, colour = g)) +
    geom_freqpoly(binwidth = binwidth, size = 1) +
    coord_cartesian(xlim = xlim)
}

t_test <- function(x1, x2) {
  test <- t.test(x1, x2)

  # use sprintf() to format t.test() results compactly
  sprintf(
    "p value: %0.3f\n[%0.2f, %0.2f]",
    test$p.value, test$conf.int[1], test$conf.int[2]
  )
}
```

If I have some simulated data, I can use these functions to compare two variables:

```
x1 <- rnorm(100, mean = 0, sd = 0.5)
x2 <- rnorm(200, mean = 0.15, sd = 0.9)

freqpoly(x1, x2)
cat(t_test(x1, x2))
#> p value: 0.003
#> [-0.38, -0.08]
```

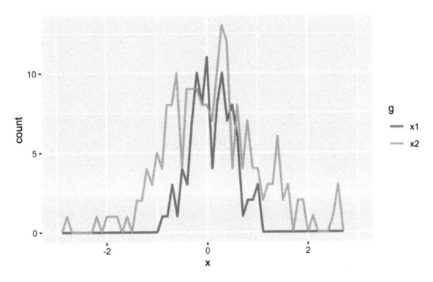

In a real analysis, you probably would've done a bunch of exploration before you ended up with these functions. I've skipped that exploration here so we can get to the app as quickly as possible. But extracting imperative code out into regular functions is an important technique for all Shiny apps: the more code you can extract out of your app, the easier it will be to understand. This is good software engineering because it helps isolate concerns: the functions outside of the app focus on the computation so that the code inside of the app can focus on responding to user actions. We'll come back to that idea again in Chapter 18.

The App

I'd like to use these two tools to quickly explore a bunch of simulations. A Shiny app is a great way to do this because it lets you avoid tediously modifying and rerunning R code. In the following, I wrap the pieces into a Shiny app, where I can interactively tweak the inputs.

Let's start with the UI. We'll come back to exactly what `fluidRow()` and `column()` do in "Multirow" on page 93; but you can guess their purpose from their names ☺. The first row has three columns for input controls (distribution 1, distribution 2, and plot controls). The second row has a wide column for the plot and a narrow column for the hypothesis test:

```
ui <- fluidPage(
  fluidRow(
    column(4,
      "Distribution 1",
      numericInput("n1", label = "n", value = 1000, min = 1),
      numericInput("mean1", label = "µ", value = 0, step = 0.1),
      numericInput("sd1", label = "σ", value = 0.5, min = 0.1, step = 0.1)
    ),
    column(4,
      "Distribution 2",
      numericInput("n2", label = "n", value = 1000, min = 1),
      numericInput("mean2", label = "µ", value = 0, step = 0.1),
      numericInput("sd2", label = "σ", value = 0.5, min = 0.1, step = 0.1)
    ),
    column(4,
      "Frequency polygon",
      numericInput("binwidth", label = "Bin width", value = 0.1, step = 0.1),
      sliderInput("range", label = "range", value = c(-3, 3), min = -5, max = 5)
    )
  ),
  fluidRow(
    column(9, plotOutput("hist")),
    column(3, verbatimTextOutput("ttest"))
  )
)
```

The server function combines calls to `freqpoly()` and `t_test()` functions after drawing from the specified distributions:

```
server <- function(input, output, session) {
  output$hist <- renderPlot({
    x1 <- rnorm(input$n1, input$mean1, input$sd1)
    x2 <- rnorm(input$n2, input$mean2, input$sd2)

    freqpoly(x1, x2, binwidth = input$binwidth, xlim = input$range)
  }, res = 96)

  output$ttest <- renderText({
    x1 <- rnorm(input$n1, input$mean1, input$sd1)
    x2 <- rnorm(input$n2, input$mean2, input$sd2)

    t_test(x1, x2)
  })
}
```

This definition of `server` and `ui` yields Figure 3-6. I recommend opening the live version and having a quick play to make sure you understand its basic operation before you continue reading.

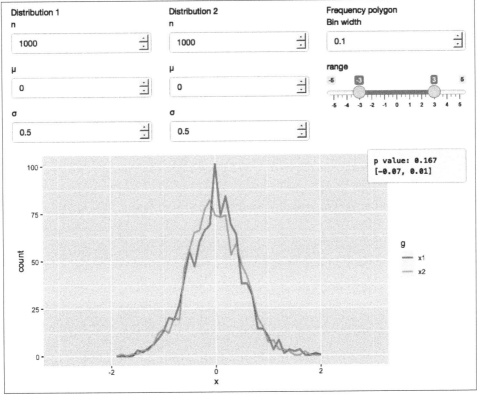

Figure 3-6. This Shiny app lets you compare two simulated distributions with a t-test and a frequency polygon. See live at https://hadley.shinyapps.io/ms-case-study-1.

The Reactive Graph

Let's start by drawing the reactive graph of this app. Shiny is smart enough to update an output only when the inputs it refers to change; it's not smart enough to only selectively run pieces of code inside an output. In other words, outputs are atomic: they're either executed or not as a whole.

For example, take this snippet from the server:

```
x1 <- rnorm(input$n1, input$mean1, input$sd1)
x2 <- rnorm(input$n2, input$mean2, input$sd2)
t_test(x1, x2)
```

As a human reading this code, you can tell that we only need to update x1 when n1, mean1, or sd1 changes, and we only need to update x2 when n2, mean2, or sd2 changes. Shiny, however, only looks at the output as a whole, so it will update both x1 and x2 every time one of n1, mean1, sd1, n2, mean2, or sd2 changes. This leads to the reactive graph shown in Figure 3-7:

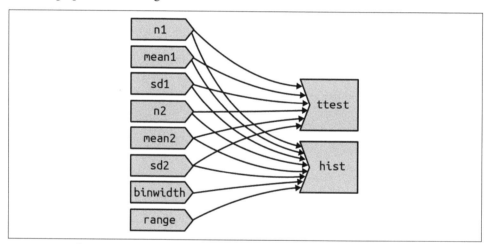

Figure 3-7. The reactive graph shows that every output depends on every input.

You'll notice that the graph is very dense: almost every input is connected directly to every output. This creates two problems:

- The app is hard to understand because there are so many connections. There are no pieces of the app that you can pull out and analyze in isolation.

- The app is inefficient because it does more work than necessary. For example, if you change the breaks of the plot, the data is recalculated; if you change the value of n1, x2 is updated (in two places!).

There's one other major flaw in the app: the frequency polygon and t-test use separate random draws. This is rather misleading, as you'd expect them to be working on the same underlying data.

Fortunately, we can fix all these problems by using reactive expressions to pull out repeated computation.

Simplifying the Graph

In the following server function, we refactor the existing code to pull out the repeated code into two new reactive expressions, x1 and x2, which simulate the data from the two distributions. To create a reactive expression, we call reactive() and assign the results to a variable. To later use the expression, we call the variable like it's a function:

```
server <- function(input, output, session) {
  x1 <- reactive(rnorm(input$n1, input$mean1, input$sd1))
  x2 <- reactive(rnorm(input$n2, input$mean2, input$sd2))

  output$hist <- renderPlot({
    freqpoly(x1(), x2(), binwidth = input$binwidth, xlim = input$range)
  }, res = 96)

  output$ttest <- renderText({
    t_test(x1(), x2())
  })
}
```

This transformation yields the substantially simpler graph shown in Figure 3-8.

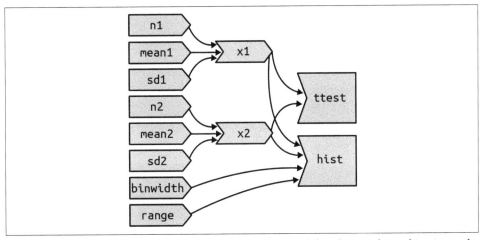

Figure 3-8. Using reactive expressions considerably simplifies the graph, making it much easier to understand.

This simpler graph makes it easier to understand the app because you can understand connected components in isolation; the values of the distribution parameters only affect the output via x1 and x2. This rewrite also makes the app much more efficient since it does much less computation. Now, when you change the binwidth or range, only the plot changes, not the underlying data.

To emphasize this modularity, Figure 3-9 draws boxes around the independent components. We'll come back to this idea in Chapter 19, when we discuss modules. Modules allow you to extract out repeated code for reuse while guaranteeing that it's isolated from everything else in the app. Modules are an extremely useful and powerful technique for more complex apps.

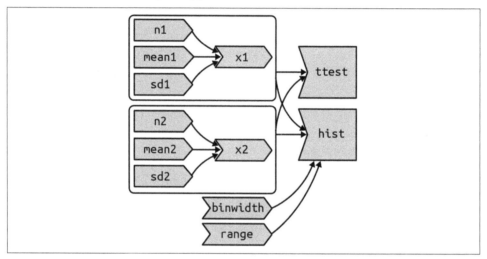

Figure 3-9. Modules enforce isolation between parts of an app.

You might be familiar with the "rule of three" of programming: whenever you copy and paste something three times, you should figure out how to reduce the duplication (typically by writing a function). This is important because it reduces the amount of duplication in your code, which makes it easier to understand and easier to update as your requirements change.

In Shiny, however, I think you should consider the rule of one: whenever you copy and paste something *once*, you should consider extracting the repeated code out into a reactive expression. The rule is stricter for Shiny because reactive expressions don't just make it easier for humans to understand the code, but they also improve Shiny's ability to efficiently rerun code.

Why Do We Need Reactive Expressions?

When you first start working with reactive code, you might wonder why we need reactive expressions. Why can't you use your existing tools—creating new variables and writing functions—for reducing duplication in code? Unfortunately, neither of these techniques work in a reactive environment.

If you try to use a variable to reduce duplication, you might write something like this:

```r
server <- function(input, output, session) {
  x1 <- rnorm(input$n1, input$mean1, input$sd1)
  x2 <- rnorm(input$n2, input$mean2, input$sd2)

  output$hist <- renderPlot({
    freqpoly(x1, x2, binwidth = input$binwidth, xlim = input$range)
  }, res = 96)

  output$ttest <- renderText({
    t_test(x1, x2)
  })
}
```

If you run this code, you'll get an error because you're attempting to access input values outside of a reactive context. Even if you didn't get that error, you'd still have a problem: x1 and x2 would only be computed once, when the session begins, not every time one of the inputs was updated.

If you try to use a function, the app will work:

```r
server <- function(input, output, session) {
  x1 <- function() rnorm(input$n1, input$mean1, input$sd1)
  x2 <- function() rnorm(input$n2, input$mean2, input$sd2)

  output$hist <- renderPlot({
    freqpoly(x1(), x2(), binwidth = input$binwidth, xlim = input$range)
  }, res = 96)

  output$ttest <- renderText({
    t_test(x1(), x2())
  })
}
```

But it has the same problem as the original code: any input will cause all outputs to be recomputed, and the t-test and the frequency polygon will be run on separate samples. Reactive expressions automatically cache their results and only update when their inputs change.[9]

While variables calculate the value only once (the porridge is too cold), and functions calculate the value every time they're called (the porridge is too hot), reactive expressions calculate the value only when it might have changed (the porridge is just right!).

Controlling Timing of Evaluation

Now that you're familiar with the basic ideas of reactivity, we'll discuss two more advanced techniques that allow you to either increase or decrease how often a reactive expression is executed. Here, I'll show how to use the basic techniques; in Chapter 15, we'll come back to their underlying implementations.

To explore the basic ideas, I'm going to simplify my simulation app. I'll use a distribution with only one parameter, and force both samples to share the same n. I'll also remove the plot controls. This yields a smaller UI object and server function:

```r
ui <- fluidPage(
  fluidRow(
    column(3,
      numericInput("lambda1", label = "lambda1", value = 3),
      numericInput("lambda2", label = "lambda2", value = 5),
      numericInput("n", label = "n", value = 1e4, min = 0)
    ),
    column(9, plotOutput("hist"))
  )
)
server <- function(input, output, session) {
  x1 <- reactive(rpois(input$n, input$lambda1))
  x2 <- reactive(rpois(input$n, input$lambda2))
  output$hist <- renderPlot({
    freqpoly(x1(), x2(), binwidth = 1, xlim = c(0, 40))
  }, res = 96)
}
```

This generates the app shown in Figure 3-10 and reactive graph shown in Figure 3-11.

---

9 If you're familiar with memoization, this is a similar idea.

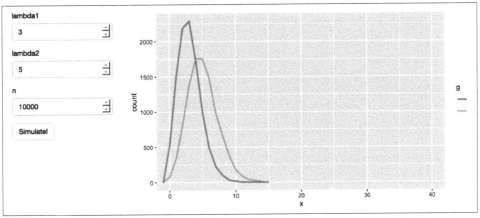

Figure 3-10. This simpler app displays a frequency polygon of random numbers drawn from two Poisson distributions. See live at https://hadley.shinyapps.io/ms-simulation-2.

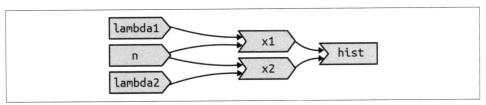

Figure 3-11. The reactive graph.

Timed Invalidation

Imagine you wanted to reinforce the fact that this is for simulated data by constantly resimulating the data so that you see an animation rather than a static plot.[10] We can increase the frequency of updates with a new function: `reactiveTimer()`.

`reactiveTimer()` is a reactive expression that has a dependency on a hidden input: the current time. You can use a `reactiveTimer()` when you want a reactive expression to invalidate itself more often than it otherwise would. For example, the following code uses an interval of 500 ms so that the plot will update twice a second. This is fast enough to remind you that you're looking at a simulation, without dizzying you with rapid changes. This change yields the reactive graph shown in Figure 3-12:

```
server <- function(input, output, session) {
  timer <- reactiveTimer(500)

  x1 <- reactive({
```

---

10 The *New York Times* used this technique particularly effectively in their article discussing how to interpret the jobs report (*https://oreil.ly/PhqSA*).

```
  timer()
  rpois(input$n, input$lambda1)
})
x2 <- reactive({
  timer()
  rpois(input$n, input$lambda2)
})

output$hist <- renderPlot({
  freqpoly(x1(), x2(), binwidth = 1, xlim = c(0, 40))
}, res = 96)
}
```

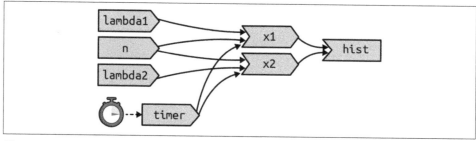

Figure 3-12. `reactiveTimer(500)` introduces a new reactive input that automatically invalidates every half a second.

Note how we use `timer()` in the reactive expressions that compute `x1()` and `x2()`: we call it but don't use the value. This lets `x1` and `x2` take a reactive dependency on `timer`, without worrying about exactly what value it returns.

On Click

In the preceding scenario, think about what would happen if the simulation code took 1 second to run. We perform the simulation every 0.5s, so Shiny would have more and more to do and would never be able to catch up. The same problem can happen if someone is rapidly clicking buttons in your app and the computation you are doing is relatively expensive. It's possible to create a big backlog of work for Shiny, and while it's working on the backlog, it can't respond to any new events. This leads to a poor user experience.

If this situation arises in your app, you might want to require the user to opt in to performing the expensive calculation by requiring them to click a button. This is a great use case for an `actionButton()`:

```
ui <- fluidPage(
  fluidRow(
    column(3,
      numericInput("lambda1", label = "lambda1", value = 3),
      numericInput("lambda2", label = "lambda2", value = 5),
```

```
      numericInput("n", label = "n", value = 1e4, min = 0),
      actionButton("simulate", "Simulate!")
    ),
    column(9, plotOutput("hist"))
  )
)
```

To use the action button, we need to learn a new tool. To see why, let's first tackle the problem using the previous approach. As we did before, we refer to simulate without using its value to take a reactive dependency on it:

```
server <- function(input, output, session) {
  x1 <- reactive({
    input$simulate
    rpois(input$n, input$lambda1)
  })
  x2 <- reactive({
    input$simulate
    rpois(input$n, input$lambda2)
  })
  output$hist <- renderPlot({
    freqpoly(x1(), x2(), binwidth = 1, xlim = c(0, 40))
  }, res = 96)
}
```

This yields the app in Figure 3-13 and reactive graph in Figure 3-14. This doesn't achieve our goal because it just introduces a new dependency: x1() and x2() will update when we click the simulate button, but they'll also continue to update when lambda1, lambda2, or n change. We want to *replace* the existing dependencies, not add to them.

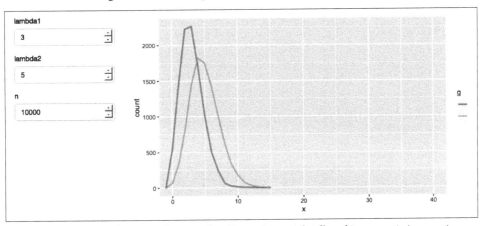

Figure 3-13. App with action button. See live at https://hadley.shinyapps.io/ms-action-button.

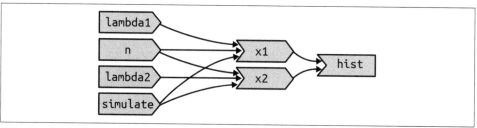

Figure 3-14. This reactive graph doesn't accomplish our goal; we've added a dependency instead of replacing the existing dependencies.

To solve this problem we need a new tool: a way to use input values without taking a reactive dependency on them. We need `eventReactive()`, which has two arguments: the first argument specifies what to take a dependency on, and the second argument specifies what to compute. That allows this app to only compute `x1()` and `x2()` when `simulate` is clicked:

```
server <- function(input, output, session) {
  x1 <- eventReactive(input$simulate, {
    rpois(input$n, input$lambda1)
  })
  x2 <- eventReactive(input$simulate, {
    rpois(input$n, input$lambda2)
  })

  output$hist <- renderPlot({
    freqpoly(x1(), x2(), binwidth = 1, xlim = c(0, 40))
  }, res = 96)
}
```

Figure 3-15 shows the new reactive graph. Note that, as desired, x1 and x2 no longer have a reactive dependency on `lambda1`, `lambda2`, and n: changing their values will not trigger computation. I left the arrows in very pale gray just to remind you that x1 and x2 continue to use the values but no longer take a reactive dependency on them.

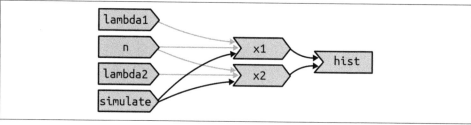

Figure 3-15. `eventReactive()` makes it possible to separate the dependencies (black arrows) from the values used to compute the result (pale-gray arrows).

Observers

So far, we've focused on what's happening inside the app. But sometimes you need to reach outside of the app and cause side effects to happen elsewhere in the world. This might be saving a file to a shared network drive, sending data to a web API, updating a database, or (most commonly) printing a debugging message to the console. These actions don't affect how your app looks, so you shouldn't use an output and a `render` function. Instead you need to use an *observer*.

There are multiple ways to create an observer, and we'll come back to them later in "Observers and Outputs" on page 226. For now, I wanted to show you how to use `observeEvent()`, because it gives you an important debugging tool when you're first learning Shiny.

`observeEvent()` is very similar to `eventReactive()`. It has two important arguments: `eventExpr` and `handlerExpr`. The first argument is the input or expression to take a dependency on; the second argument is the code that will be run. For example, the following modification to `server()` means that every time that `name` is updated, a message will be sent to the console:

```r
ui <- fluidPage(
  textInput("name", "What's your name?"),
  textOutput("greeting")
)

server <- function(input, output, session) {
  string <- reactive(paste0("Hello ", input$name, "!"))

  output$greeting <- renderText(string())
  observeEvent(input$name, {
    message("Greeting performed")
  })
}
```

There are two important differences between `observeEvent()` and `eventReactive()`:

- You don't assign the result of `observeEvent()` to a variable.
- As a result, you can't refer to it from other reactive consumers.

Observers and outputs are closely related. You can think of outputs as having a special side effect: updating the HTML in the user's browser. To emphasize this closeness, we'll draw them the same way in the reactive graph. This yields the reactive graph shown in Figure 3-16.

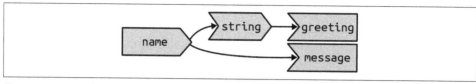

Figure 3-16. In the reactive graph, an observer looks the same as an output.

Summary

This chapter should have improved your understanding of the backend of Shiny apps, the `server()` code that responds to user actions. You've also taken the first steps in mastering the reactive programming paradigm that underpins Shiny. What you've learned here will carry you a long way; we'll come back to the underlying theory in Chapter 13. Reactivity is extremely powerful, but it is also very different from the imperative style of R programming that you're most used to. Don't be surprised if it takes a while for all the consequences to sink in.

This chapter concludes our overview of the foundations of Shiny. The next chapter will help you practice the material you've seen so far by creating a bigger Shiny app designed to support a data analysis.

## CHAPTER 4

Case Study: ER Injuries

Introduction

I've introduced you to a bunch of new concepts in the last three chapters. So to help them sink in, we'll now walk through a richer Shiny app that explores a fun dataset and pulls together many of the ideas that you've seen so far. We'll start by doing a little data analysis outside of Shiny, then turn it into an app, starting simply, then progressively layering on more detail.

In this chapter, we'll supplement Shiny with vroom (for fast file reading) and the tidyverse (for general data analysis):

```
library(shiny)
library(vroom)
library(tidyverse)
```

The Data

We're going to explore data from the National Electronic Injury Surveillance System (NEISS), collected by the Consumer Product Safety Commission. This is a long-term study that records all accidents seen in a representative sample of hospitals in the United States. It's an interesting dataset to explore because everyone is already familiar with the domain, and each observation is accompanied by a short narrative that explains how the accident occurred. You can find out more about this dataset on GitHub (*https://github.com/hadley/neiss*).

In this chapter, I'm going to focus on just the data from 2017. This keeps the data small enough (~10 MB) that it's easy to store in Git (along with the rest of the book), which means we don't need to think about sophisticated strategies for importing the data quickly (we'll come back to those later in the book). You can see the code I used to create the extract for this chapter on GitHub (*https://oreil.ly/ERRCh*).

If you want to get the data on to your own computer, run this code:

```
dir.create("neiss")
#> Warning in dir.create("neiss"): 'neiss' already exists
download <- function(name) {
  url <- "https://github.com/hadley/mastering-shiny/raw/master/neiss/"
  download.file(paste0(url, name), paste0("neiss/", name), quiet = TRUE)
}
download("injuries.tsv.gz")
download("population.tsv")
download("products.tsv")
```

The main dataset we'll use is `injuries`, which contains around 250,000 observations:

```
injuries <- vroom::vroom("neiss/injuries.tsv.gz")
injuries
#> # A tibble: 255,064 x 10
#>   trmt_date  age sex   race  body_part diag       location   prod_code weight
#>   <date>     <dbl> <chr> <chr> <chr>     <chr>      <chr>          <dbl> <dbl>
#> 1 2017-01-01    71 male  white Upper Tru… Contusion… Other Publ…     1807  77.7
#> 2 2017-01-01    16 male  white Lower Arm Burns, Th… Home             676  77.7
#> 3 2017-01-01    58 male  white Upper Tru… Contusion… Home             649  77.7
#> 4 2017-01-01    21 male  white Lower Tru… Strain, S… Home            4076  77.7
#> 5 2017-01-01    54 male  white Head      Inter Org… Other Publ…     1807  77.7
#> 6 2017-01-01    21 male  white Hand      Fracture   Home            1884  77.7
#> # … with 255,058 more rows, and 1 more variable: narrative <chr>
```

Each row represents a single accident with 10 variables:

trmt_date

The date the person was seen in the hospital (not when the accident occurred).

age, sex, *and* race

Demographic information about the person who experienced the accident.

body_part

The location of the injury on the body (like ankle or ear); location is the place where the accident occurred (like home or school).

diag

The basic diagnosis of the injury (like fracture or laceration).

prod_code

The primary product associated with the injury.

weight

The statistical weight giving the estimated number of people who would suffer this injury if this dataset was scaled to the entire population of the US.

narrative

A brief story about how the accident occurred.

We'll pair it with two other data frames for additional context: products lets us look up the product name from the product code, and population tells us the total US population in 2017 for each combination of age and sex:

```
products <- vroom::vroom("neiss/products.tsv")
products
#> # A tibble: 38 x 2
#>    prod_code title
#>        <dbl> <chr>
#> 1        464 knives, not elsewhere classified
#> 2        474 tableware and accessories
#> 3        604 desks, chests, bureaus or buffets
#> 4        611 bathtubs or showers
#> 5        649 toilets
#> 6        676 rugs or carpets, not specified
#> # … with 32 more rows

population <- vroom::vroom("neiss/population.tsv")
population
#> # A tibble: 170 x 3
#>     age sex      population
#>   <dbl> <chr>         <dbl>
#> 1     0 female      1924145
#> 2     0 male        2015150
#> 3     1 female      1943534
#> 4     1 male        2031718
#> 5     2 female      1965150
#> 6     2 male        2056625
#> # … with 164 more rows
```

Exploration

Before we create the app, let's explore the data a little. We'll start by looking at a product with an interesting story: 649, "toilets." First we'll pull out the injuries associated with this product:

```
selected <- injuries %>% filter(prod_code == 649)
nrow(selected)
#> [1] 2993
```

Next we'll perform some basic summaries looking at the location, body part, and diagnosis of toilet-related injuries. Note that I weight by the weight variable so that the counts can be interpreted as estimated total injuries across the whole US:

```
selected %>% count(location, wt = weight, sort = TRUE)
#> # A tibble: 6 x 2
#>   location                  n
#>   <chr>                 <dbl>
#> 1 Home                 99603.
#> 2 Other Public Property 18663.
#> 3 Unknown              16267.
```

```
#> 4 School                    659.
#> 5 Street Or Highway          16.2
#> 6 Sports Or Recreation Place 14.8

selected %>% count(body_part, wt = weight, sort = TRUE)
#> # A tibble: 24 x 2
#>    body_part          n
#>    <chr>          <dbl>
#> 1 Head           31370.
#> 2 Lower Trunk    26855.
#> 3 Face           13016.
#> 4 Upper Trunk    12508.
#> 5 Knee            6968.
#> 6 N.S./Unk        6741.
#> # … with 18 more rows

selected %>% count(diag, wt = weight, sort = TRUE)
#> # A tibble: 20 x 2
#>    diag                     n
#>    <chr>                <dbl>
#> 1 Other Or Not Stated   32897.
#> 2 Contusion Or Abrasion 22493.
#> 3 Inter Organ Injury    21525.
#> 4 Fracture              21497.
#> 5 Laceration            18734.
#> 6 Strain, Sprain         7609.
#> # … with 14 more rows
```

As you might expect, injuries involving toilets most often occur at home. The most
common body parts involved possibly suggest that these are falls (since the head and
face are not usually involved in routine toilet usage), and the diagnoses seem rather
varied.

We can also explore the pattern across age and sex. We have enough data here that a
table is not that useful, and so I make a plot, as seen in Figure 4-1, that makes the
patterns more obvious:

```
summary <- selected %>%
  count(age, sex, wt = weight)
summary
#> # A tibble: 208 x 3
#>    age sex         n
#>    <dbl> <chr>   <dbl>
#> 1     0 female   4.76
#> 2     0 male    14.3
#> 3     1 female 253.
#> 4     1 male   231.
#> 5     2 female 438.
#> 6     2 male   632.
#> # … with 202 more rows

summary %>%
```

```
ggplot(aes(age, n, colour = sex)) +
geom_line() +
labs(y = "Estimated number of injuries")
```

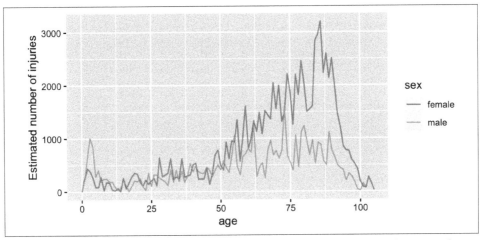

Figure 4-1. Estimated number of injuries caused by toilets, broken down by age and sex.

We see a spike for young boys peaking at age 3, and then an increase (particularly for women) starting around middle age, and a gradual decline after age 80. I suspect the peak is because boys usually use the toilet standing up, and the increase for women is due to osteoporosis (i.e., I suspect women and men have injuries at the same rate, but more women end up in the ER because they are at higher risk of fractures).

One problem with interpreting this pattern is that we know that there are fewer older people than younger people, so the population available to be injured is smaller. We can control for this by comparing the number of people injured with the total population and calculating an injury rate. Here I use a rate per 10,000:

```
summary <- selected %>%
  count(age, sex, wt = weight) %>%
  left_join(population, by = c("age", "sex")) %>%
  mutate(rate = n / population * 1e4)

summary
#> # A tibble: 208 x 5
#>     age sex         n population   rate
#>   <dbl> <chr>   <dbl>      <dbl>  <dbl>
#> 1     0 female   4.76    1924145 0.0247
#> 2     0 male    14.3     2015150 0.0708
#> 3     1 female 253.      1943534 1.30
#> 4     1 male   231.      2031718 1.14
#> 5     2 female 438.      1965150 2.23
#> 6     2 male   632.      2056625 3.07
#> # … with 202 more rows
```

Plotting the rate, as shown in Figure 4-2, yields a strikingly different trend after age 50: the difference between men and women is much smaller, and we no longer see a decrease. This is because women tend to live longer than men, so at older ages there are simply more women alive to be injured by toilets:

```
summary %>%
  ggplot(aes(age, rate, colour = sex)) +
  geom_line(na.rm = TRUE) +
  labs(y = "Injuries per 10,000 people")
```

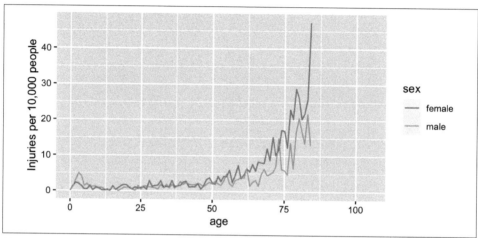

Figure 4-2. Estimated rate of injuries per 10,000 people, broken down by age and sex.

(Note that the rates only go up to age 80 because I couldn't find population data for ages over 80.)

Finally, we can look at some of the narratives. Browsing through these is an informal way to check our hypotheses and generate new ideas for further exploration. Here I pull out a random sample of 10:

```
selected %>%
  sample_n(10) %>%
  pull(narrative)
#>  [1] "68YOF STRAINED KNEE MOVING FROM TOILET TO POWER CHAIR AT HOME.  DX:...
#>  [2] "97YOM LWR BACK PAIN - MISSED TOILET SEAT, FELL FLOOR AT NH"
#>  [3] "54 YOF DX ALCOHOL INTOXICATION - PT STATES SHE FELL OFF TOILET."
#>  [4] "85YOF-STAFF AT NH STATES PT WAS TRANSITIONIN TO TOILET FROM WHEELCH...
#>  [5] "FOREHEAD LACERATION. 64 YOM FELL AND HIT HIS HEAD ON TOILET."
#>  [6] "70YOM-STAFF STATES PT FELL OFF TOILET ONTO CONCRETE FLOOR AT *** AR...
#>  [7] "40YOF WAS INTOXICATED AND FELL OFF THE TOILET STRUCK HEAD ON THE WA...
#>  [8] "66 Y/O F FELL FROM COMMODE ONTO FLOOR AND FRACTURED CLAVICLE"
#>  [9] "25YOF SYNCOPAL EPS W ON TOILET FELL HIT RS OF HEAD REPORTLY LOC UNK...
#> [10] "4 YO M W/LAC TO FOREHEAD SLIPPED IN BATHROOM HIT ON TOILET FLUSH HA...
```

Having done this exploration for one product, it would be very nice if we could easily do it for other products, without having to retype the code. So let's make a Shiny app!

Prototype

When building a complex app, I strongly recommend starting as simple as possible so that you can confirm that the basic mechanics work before you start doing something more complicated. Here I'll start with one input (the product code), three tables, and one plot.

When designing a first prototype, the challenge is in making it "as simple *as possible.*" There's a tension between getting the basics working quickly and planning for the future of the app. Either extreme can be bad: if you design too narrowly, you'll spend a lot of time later reworking your app; if you design too rigorously, you'll spend a bunch of time writing code that later ends up on the cutting room floor. To help get the balance right, I often do a few pencil-and-paper sketches to rapidly explore the UI and reactive graph before committing to code.

Here I decided to have 1 row for the inputs (accepting that I'm probably going to add more inputs before this app is done), 1 row for all three tables (giving each table 4 columns, 1/3 of the 12-column width), and then 1 row for the plot:

```r
prod_codes <- setNames(products$prod_code, products$title)

ui <- fluidPage(
  fluidRow(
    column(6,
      selectInput("code", "Product", choices = prod_codes)
    )
  ),
  fluidRow(
    column(4, tableOutput("diag")),
    column(4, tableOutput("body_part")),
    column(4, tableOutput("location"))
  ),
  fluidRow(
    column(12, plotOutput("age_sex"))
  )
)
```

We haven't talked about fluidRow() and column() yet, but you should be able to guess what they do from the context, and we'll come back to talk about them in "Multirow" on page 93. Also note the use of setNames() in the selectInput() choices: this shows the product name in the UI and returns the product code to the server.

The server function is relatively straightforward. I first convert the static `selected` and `summary` variables to reactive expressions. This is a reasonable general pattern: you create variables in your data analysis to decompose the analysis into steps and to avoid recomputing things multiple times, and reactive expressions play the same role in Shiny apps.

Often it's a good idea to spend a little time cleaning up your analysis code before you start your Shiny app, so you can think about these problems in regular R code before you add the additional complexity of reactivity:

```r
server <- function(input, output, session) {
  selected <- reactive(injuries %>% filter(prod_code == input$code))

  output$diag <- renderTable(
    selected() %>% count(diag, wt = weight, sort = TRUE)
  )
  output$body_part <- renderTable(
    selected() %>% count(body_part, wt = weight, sort = TRUE)
  )
  output$location <- renderTable(
    selected() %>% count(location, wt = weight, sort = TRUE)
  )

  summary <- reactive({
    selected() %>%
      count(age, sex, wt = weight) %>%
      left_join(population, by = c("age", "sex")) %>%
      mutate(rate = n / population * 1e4)
  })

  output$age_sex <- renderPlot({
    summary() %>%
      ggplot(aes(age, n, colour = sex)) +
      geom_line() +
      labs(y = "Estimated number of injuries")
  }, res = 96)
}
```

Note that creating the `summary` reactive isn't strictly necessary here, as it's only used by a single reactive consumer. But it's good practice to keep computing and plotting separate as it makes the flow of the app easier to understand and will make it easier to generalize in the future.

A screenshot of the resulting app is shown in Figure 4-3. You can view the source code on GitHub (*https://oreil.ly/L7xcN*).

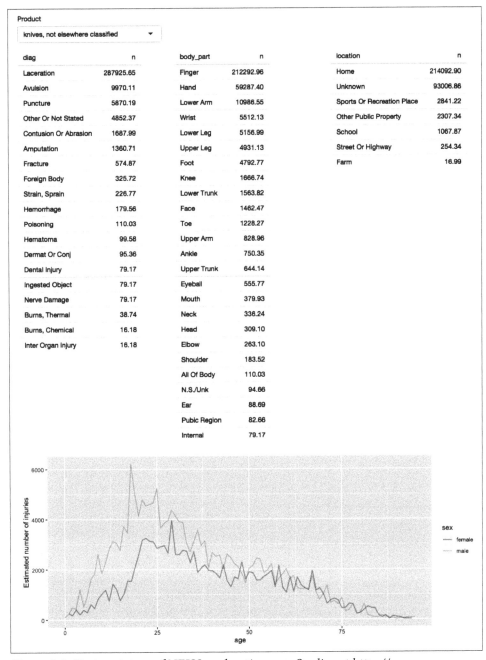

Product

knives, not elsewhere classified ▾

diag	n		body_part	n		location	n
Laceration	287925.65		Finger	212292.96		Home	214092.90
Avulsion	9970.11		Hand	59287.40		Unknown	93006.86
Puncture	5870.19		Lower Arm	10986.55		Sports Or Recreation Place	2841.22
Other Or Not Stated	4852.37		Wrist	5512.13		Other Public Property	2307.34
Contusion Or Abrasion	1687.99		Lower Leg	5156.99		School	1067.87
Amputation	1360.71		Upper Leg	4931.13		Street Or Highway	254.34
Fracture	574.87		Foot	4792.77		Farm	16.99
Foreign Body	325.72		Knee	1666.74			
Strain, Sprain	226.77		Lower Trunk	1563.82			
Hemorrhage	179.56		Face	1462.47			
Poisoning	110.03		Toe	1228.27			
Hematoma	99.58		Upper Arm	828.96			
Dermat Or Conj	95.36		Ankle	750.35			
Dental Injury	79.17		Upper Trunk	644.14			
Ingested Object	79.17		Eyeball	555.77			
Nerve Damage	79.17		Mouth	379.93			
Burns, Thermal	38.74		Neck	336.24			
Burns, Chemical	16.18		Head	309.10			
Inter Organ Injury	16.18		Elbow	263.10			
			Shoulder	183.52			
			All Of Body	110.03			
			N.S./Unk	94.66			
			Ear	88.69			
			Pubic Region	82.66			
			Internal	79.17			

Figure 4-3. First prototype of NEISS exploration app. See live at https://hadley.shinyapps.io/ms-prototype.

Polish Tables

Now that we have the basic components in place and working, we can progressively improve our app. The first problem with this app is that it shows a lot of information in the tables, where we probably just want the highlights. To fix this we need to first figure out how to truncate the tables. I've chosen to do that with a combination of forcats functions: I convert the variable to a factor, order by the frequency of the levels, and then lump together all levels after the top five:

```
injuries %>%
  mutate(diag = fct_lump(fct_infreq(diag), n = 5)) %>%
  group_by(diag) %>%
  summarise(n = as.integer(sum(weight)))
#> # A tibble: 6 x 2
#>   diag                        n
#> * <fct>                   <int>
#> 1 Other Or Not Stated   1806436
#> 2 Fracture              1558961
#> 3 Laceration            1432407
#> 4 Strain, Sprain        1432556
#> 5 Contusion Or Abrasion 1451987
#> 6 Other                 1929147
```

Because I knew how to do it, I wrote a little function to automate this for any variable. The details aren't really important here, but we'll come back to them in Chapter 12. You could also solve the problem with copy and paste, so don't worry if the code looks totally foreign:

```
count_top <- function(df, var, n = 5) {
  df %>%
    mutate({{ var }} := fct_lump(fct_infreq({{ var }}), n = n)) %>%
    group_by({{ var }}) %>%
    summarise(n = as.integer(sum(weight)))
}
```

I then use this in the server function:

```
output$diag <- renderTable(count_top(selected(), diag), width = "100%")
output$body_part <- renderTable(count_top(selected(), body_part), width = "100%")
output$location <- renderTable(count_top(selected(), location), width = "100%")
```

I made one other change to improve the aesthetics of the app: I forced all tables to take up the maximum width (i.e., fill the column that they appear in). This makes the output more aesthetically pleasing because it reduces the amount of incidental variation.

A screenshot of the resulting app is shown in Figure 4-4. You can view the source code on GitHub (*https://oreil.ly/Mya71*).

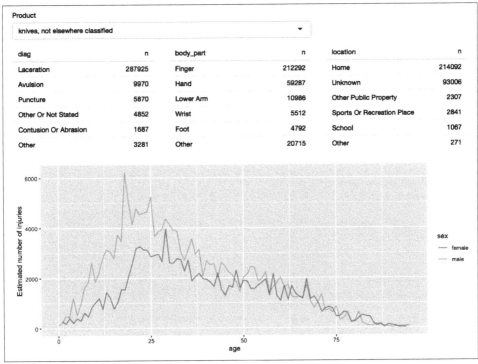

Figure 4-4. The second iteration of the app improves the display by only showing the most frequent rows in the summary tables. See live at https://hadley.shinyapps.io/ms-polish-tables.

Rate Versus Count

So far, we're displaying only a single plot, but we'd like to give the user the choice between visualizing the number of injuries or the population-standardized rate. First I add a control to the UI. Here I've chosen to use a `selectInput()` because it makes both states explicit, and it would be easy to add new states in the future:

```
fluidRow(
  column(8,
    selectInput("code", "Product",
      choices = setNames(products$prod_code, products$title),
      width = "100%"
    )
  ),
  column(2, selectInput("y", "Y axis", c("rate", "count")))
),
```

(I default to `rate` because I think it's safer; you don't need to understand the popula-tion distribution in order to correctly interpret the plot.)

Then I condition on that input when generating the plot:

```
output$age_sex <- renderPlot({
  if (input$y == "count") {
    summary() %>%
      ggplot(aes(age, n, colour = sex)) +
      geom_line() +
      labs(y = "Estimated number of injuries")
  } else {
    summary() %>%
      ggplot(aes(age, rate, colour = sex)) +
      geom_line(na.rm = TRUE) +
      labs(y = "Injuries per 10,000 people")
  }
}, res = 96)
```

A screenshot of the resulting app is shown in Figure 4-5. You can view the source code on GitHub (*https://oreil.ly/3AwYf*).

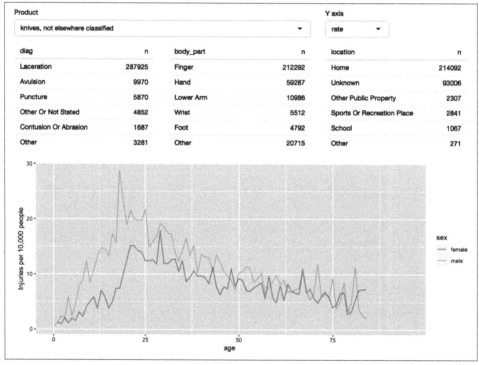

Figure 4-5. In this iteration, we give the user the ability to switch between displaying the count and the population standardized rate on the y-axis. See live at https:// hadley.shinyapps.io/ms-rate-vs-count.

Narrative

Finally, I want to provide some way to access the narratives because they are so interesting, and they give an informal way to cross-check the hypotheses you come up with when looking at the plots. In the R code, I sample multiple narratives at once, but there's no reason to do that in an app where you can explore interactively.

There are two parts to the solution. First we add a new row to the bottom of the UI. I use an action button to trigger a new story, and put the narrative in a `textOutput()`:

```
fluidRow(
  column(2, actionButton("story", "Tell me a story")),
  column(10, textOutput("narrative"))
)
```

A screenshot of the resulting app is shown in Figure 4-6. You can view the source code on GitHub (*https://oreil.ly/dsqIH*).

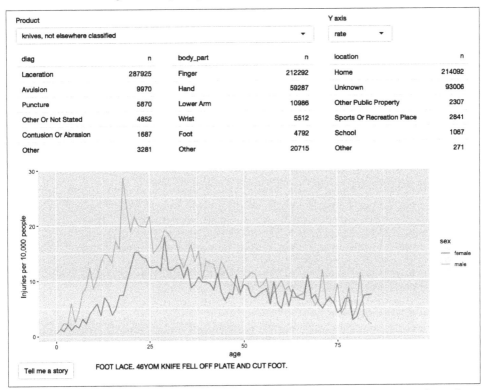

Figure 4-6. The final iteration adds the ability to pull out a random narrative from the selected rows. See live at https://hadley.shinyapps.io/ms-narrative.

I then use `eventReactive()` to create a reactive that only updates when the button is clicked or the underlying data changes:

```
narrative_sample <- eventReactive(
  list(input$story, selected()),
  selected() %>% pull(narrative) %>% sample(1)
)
output$narrative <- renderText(narrative_sample())
```

Exercises

1. Draw the reactive graph for each app.

2. What happens if you flip `fct_infreq()` and `fct_lump()` in the code that reduces the summary tables?

3. Add an input control that lets the user decide how many rows to show in the summary tables.

4. Provide a way to step through every narrative systematically with forward and backward buttons.

   Advanced: Make the list of narratives "circular" so that advancing forward from the last narrative takes you to the first.

Summary

Now that you have the basics of Shiny apps under your belt, the following seven chapters will give you a grab bag of important techniques. Once you've read the next chapter on workflow, I recommend skimming the remaining chapters so you get a good sense of what they cover, then dip your toes back in as you need the techniques for an app.

## PART II

Shiny in Action

The following chapters give you a grab bag of useful techniques. I think everyone should start with Chapter 5, because it gives you important tools for developing and debugging apps and getting help when you're stuck. After that, there's no prescribed order and relatively few connections between the chapters: I'd suggest quickly skimming to get the lay of the land (and so you might remember these tools if related problems crop up in the future) and otherwise only deeply reading the bits that you currently need. Here's a quick rundown of the main topics:

- Chapter 6 details the various ways you can layout input and output components on a page and how you can customize their appearance with themes.

- Chapter 7 shows you how to add direct interaction to your plot and how to display images generated in other ways.

- Chapter 8 covers a family of techniques (inline errors, notifications, progress bars, dialog boxes, etc.) for giving feedback to your users while your app runs.

- Chapter 9 discusses how to transfer files to and from your app.

- Chapter 10 shows you how to dynamically modify your app's user interface while it runs.

- Chapter 11 shows how to record app state in such a way that your users can bookmark it.

- Chapter 12 shows you how to allow users to select variables when working with tidyverse packages.

Let's begin by working on your workflow for developing apps.

## CHAPTER 5

Workflow

If you're going to be writing a lot of Shiny apps (and since you're reading this book, I hope you will be!), it's worth investing some time in your basic workflow. Improving workflow is a good place to invest time because it tends to pay great dividends in the long run. It doesn't just increase the proportion of your time spent writing R code, but because you see the results more quickly, it makes the process of writing Shiny apps more enjoyable and helps your skills improve more quickly.

The goal of this chapter is to help you improve three important Shiny workflows:

- The basic development cycle of creating apps, making changes, and experimenting with the results.
- Debugging, the workflow where you figure out what's gone wrong with your code and then brainstorm solutions to fix it.
- Writing reprexes, self-contained chunks of code that illustrate a problem. Reprexes are a powerful debugging technique, and they are essential if you want to get help from someone else.

Development Workflow

The goal of optimizing your development workflow is to reduce the time between making a change and seeing the outcome. The faster you can iterate, the faster you can experiment and the faster you can become a better Shiny developer. There are two main workflows to optimize here: creating an app for the first time and speeding up the iterative cycle of tweaking code and trying out the results.

Creating the App

You will start every app with the same six lines of R code:

```
library(shiny)
ui <- fluidPage(
)
server <- function(input, output, session) {
}
shinyApp(ui, server)
```

You'll likely quickly get sick of typing that code in, so RStudio provides a couple of shortcuts:

- If you already have your future *app.R* open, type **shinyapp**, then press Shift+Tab to insert the Shiny app snippet.[1]
- If you want to start a new project,[2] go to the File menu, select New Project, then select Shiny Web Application, as in Figure 5-1.

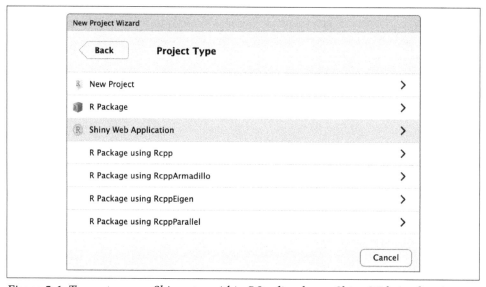

Figure 5-1. To create a new Shiny app within RStudio, choose Shiny Web Application as the project type.

---

1 Snippets (*https://oreil.ly/LDRpo*) are text macros that you can use to insert common code fragments. If you enjoy using snippets, make sure to check the collection of Shiny-specific snippets put together by ThinkR (*https://oreil.ly/kXGFY*).

2 A project is a self-contained directory that is isolated from the other projects that you're working on. If you use RStudio, but don't currently use projects, I highly recommend reading about the project-oriented lifestyle (*https://oreil.ly/px2Gi*).

You might think it's not worthwhile to learn these shortcuts because you'll only create an app or two a day, but creating simple apps is a great way to check that you have the basic concepts down before you start on a bigger project, and they're a great tool for debugging.

Seeing Your Changes

At most, you'll *create* a few apps a day, but you'll *run* apps hundreds of times, so mastering the development workflow is particularly important. The first way to reduce your iteration time is to avoid clicking on the Run App button and instead learn the keyboard shortcut Cmd/Ctrl+Shift+Enter. This gives you the following development workflow:

1. Write some code.
2. Launch the app with Cmd/Ctrl+Shift+Enter.
3. Interactively experiment with the app.
4. Close the app.
5. Go to 1.

Another way to increase your iteration speed still further is to turn autoreload on and run the app in a background job (*https://oreil.ly/jtX8d*). With this workflow, as soon as you save a file, your app will relaunch: no need to close and restart. This leads to an even faster workflow:

1. Write some code, and press Cmd/Ctrl+S to save the file.
2. Interactively experiment.
3. Go to 1.

The chief disadvantage of this technique is that it's considerably harder to debug because the app is running in a separate process.

As your app gets bigger and bigger, you'll find that the "interactively experiment" step starts to become onerous. It's too hard to remember to recheck every component of your app that you might have affected with your changes. Later, in Chapter 21, you'll learn the tools of automated testing, which allows you to turn the interactive experiments you're running into automated code. This lets you run the tests more quickly (because they're automated) and means that you can't forget to run an important test. It requires some initial investment to develop the tests, but the investment pays off handsomely for large apps.

Controlling the View

By default, when you run the app, it will appear in a pop-out window. There are two other options that you can choose from the Run App drop-down, as shown in Figure 5-2:

- Run in Viewer Pane opens the app in the viewer pane (usually located on the right-hand side of the IDE). It's useful for smaller apps because you can see it at the same time as you run your app code.

- Run External opens the app in your usual web browser. It's useful for larger apps and when you want to see what your app looks like in the context that most users will experience it.

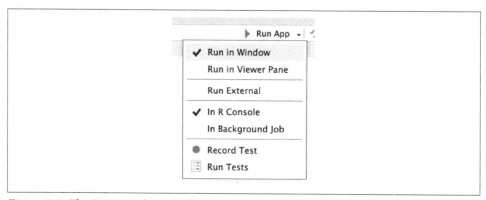

Figure 5-2. The Run App button allows you to choose how the running app will be displayed.

Debugging

When you start writing apps, it is almost guaranteed that something will go wrong. The cause of most bugs is a mismatch between your mental model of Shiny and what Shiny actually does. As you read this book, your mental model will improve so that you make fewer mistakes, and when you do make one, it's easier to spot the problem. However, it takes years of experience in any language before you can reliably write code that works the first time. This means you need to develop a robust workflow for identifying and fixing mistakes. Here we'll focus on the challenges specific to Shiny apps; if you're new to debugging in R, start with watching the "Object of type `closure` is not subsettable" (*https://oreil.ly/QXvtt*) keynote Jenny Bryan gave at rstudio::conf(2020).

There are three main cases of problems that we'll discuss:

- You get an unexpected error. This is the easiest case, because you'll get a trace-back, which allows you to figure out exactly where the error occurred. Once you've identified the problem, you'll need to systematically test your assumptions until you find a difference between your expectations and reality. The interactive debugger is a powerful assistant for this process.
- You don't get any errors, but some value is incorrect. Here, you'll need to use the interactive debugger along with your investigative skills to track down the root cause.
- All the values are correct, but they're not updated when you expect. This is the most challenging problem because it's unique to Shiny, so you can't take advantage of your existing R debugging skills.

It's frustrating when these situations arise, but you can turn them into opportunities to practice your debugging skills.

We'll come back to another important technique—making a minimal reproducible example—in the next section. Creating a minimal example is crucial if you get stuck and need to get help from someone else. But creating a minimal example is also a profoundly important skill when debugging your own code. Typically you have a lot of code that works just fine and a very small amount of code that's causing problems. If you can narrow in on the problematic code by removing the code that works, you'll be able to iterate on a solution much more quickly. This is a technique that I use every day.

Reading Tracebacks

In R, every error is accompanied by a *traceback*, or call stack, which literally traces back through the sequence of calls that lead to the error. For example, take this simple sequence of calls: f() calls g() calls h(), which calls the multiplication operator:

```
f <- function(x) g(x)
g <- function(x) h(x)
h <- function(x) x * 2
```

If this code errors, as follows:

```
f("a")
#> Error in x * 2: non-numeric argument to binary operator
```

you can call traceback() to find the sequence of calls that led to the problem:

```
traceback()
#> 3: h(x)
#> 2: g(x)
#> 1: f("a")
```

I think it's easiest to understand the traceback by flipping it upside down:

```
1: f("a")
2: g(x)
3: h(x)
```

This now tells you the sequence of calls that led to the error: f() called g() called h() (which errors).

Tracebacks in Shiny

Unfortunately, you can't use traceback() in Shiny because you can't run code while an app is running. Instead, Shiny will automatically print the traceback for you. For example, take this simple app using the f() function I defined previously:

```
library(shiny)

f <- function(x) g(x)
g <- function(x) h(x)
h <- function(x) x * 2

ui <- fluidPage(
  selectInput("n", "N", 1:10),
  plotOutput("plot")
)
server <- function(input, output, session) {
  output$plot <- renderPlot({
    n <- f(input$n)
    plot(head(cars, n))
  }, res = 96)
}
shinyApp(ui, server)
```

If you run this app, you'll see an error message in the app and a traceback in the console:

```
Error in *: non-numeric argument to binary operator
  169: g [app.R#4]
  168: f [app.R#3]
  167: renderPlot [app.R#13]
  165: func
  125: drawPlot
  111: <reactive:plotObj>
   95: drawReactive
   82: renderFunc
   81: output$plot
    1: runApp
```

To understand what's going on, we again start by flipping it upside down, so you can see the sequence of calls in the order they appear:

```
Error in *: non-numeric argument to binary operator
    1: runApp
   81: output$plot
```

```
 82: renderFunc
 95: drawReactive
111: <reactive:plotObj>
125: drawPlot
165: func
167: renderPlot [app.R#13]
168: f [app.R#3]
169: g [app.R#4]
```

There are three basic parts to the call stack:

- The first few calls start the app. In this case you just see runApp(), but depending on how you start the app, you might see something more complicated. For example, if you called source() to run the app, you might see this:

  ```
  1: source
  3: print.shiny.appobj
  5: runApp
  ```

  In general, you can ignore anything before the first runApp(); this is just the setup code to get the app running.

- Next, you'll see some internal Shiny code in charge of calling the reactive expression:

  ```
   81: output$plot
   82: renderFunc
   95: drawReactive
  111: <reactive:plotObj>
  125: drawPlot
  165: func
  ```

  Here, spotting output$plot is really important: that tells which of your reactives (plot) is causing the error. The next few functions are internal, and you can ignore them.

- Finally, at the very bottom, you'll see the code that you have written:

  ```
  167: renderPlot [app.R#13]
  168: f [app.R#3]
  169: g [app.R#4]
  ```

  This is the code called inside of renderPlot(). You can tell you should pay attention here because of the filepath and line number; this lets you know that it's your code.

If you get an error in your app but don't see a traceback, make sure that you're running the app using Cmd/Ctrl+Shift+Enter (or, if not in RStudio, calling runApp()) and that you've saved the file that you're running it from. Other ways of running the app don't always capture the information necessary to make a traceback.

The Interactive Debugger

Once you've located the source of the error and want to figure out what's causing it, the most powerful tool you have at your disposal is the *interactive debugger*. The debugger pauses execution and gives you an interactive R console where you can run any code to figure out what's gone wrong. There are two ways to launch the debugger:

- Add a call to browser() in your source code. This is the standard R way of launching the interactive debugger and will work however you're running Shiny.

  The other advantage of browser() is that because it's R code, you can make it conditional by combining it with an if statement. This allows you to launch the debugger only for problematic inputs:

  ```
  if (input$value == "a") {
    browser()
  }
  # Or maybe
  if (my_reactive() < 0) {
    browser()
  }
  ```

- Add an RStudio breakpoint by clicking to the left of the line number. You can remove the breakpoint by clicking on the red circle:

  ```
  23 ▾ server <- function(input, output, session) {
  24 ▾   territory <- reactive({
  ● 25       req(input$territory)
  26 ▾       if (input$territory == "NA") {
  ```

  The advantage of breakpoints is that they're not code, so you never have to worry about accidentally checking them into your version control system.

If you're using RStudio, the toolbar in Figure 5-3 will appear at the top of the console when you're in the debugger. The toolbar is an easy way to remember the debugging commands that are now available to you. They're also available outside of RStudio; you'll just need to remember the one-letter command to activate them. The three most useful commands are:

Next (press n)
Executes the next step in the function. Note that if you have a variable named n, you'll need to use print(n) to display its value.

Continue (press c)
Leaves interactive debugging and continues regular execution of the function. This is useful if you've fixed the bad state and want to check that the function proceeds correctly.

Stop (press Q*)*

Stops debugging, terminates the function, and returns to the global workspace. Use this once you've figured out where the problem is and you're ready to fix it and reload the code.

Figure 5-3. RStudio's debugging toolbar.

As well as stepping through the code line by line using these tools, you'll also write and run a *bunch* of interactive code to track down what's going wrong. Debugging is the process of systematically comparing your expectations to reality until you find the mismatch. If you're new to debugging in R, you might want to read the Debugging chapter (*https://oreil.ly/5jiRp*) of "Advanced R" to learn some general techniques.

Case Study

When you have eliminated the impossible, whatever remains, however improbable, must be the truth.

—Sherlock Holmes in Arthur Conan Doyle's *The Sign of the Four*

To demonstrate the basic debugging approach, I'll show you a little problem I encountered when writing "Hierarchical Select Boxes" on page 156. I'll first show you the basic context, then you'll see a problem I resolved without interactive debugging tools—a problem that required interactive debugging—and discover a final surprise.

The initial goal is pretty simple: I have a dataset of sales, and I want to filter it by territory. Here's what the data looks like:

```
sales <- readr::read_csv("sales-dashboard/sales_data_sample.csv")
sales <- sales[c(
  "TERRITORY", "ORDERDATE", "ORDERNUMBER", "PRODUCTCODE",
  "QUANTITYORDERED", "PRICEEACH"
)]
sales
#> # A tibble: 2,823 x 6
#>   TERRITORY ORDERDATE      ORDERNUMBER PRODUCTCODE QUANTITYORDERED PRICEEACH
#>   <chr>     <chr>                <dbl> <chr>                 <dbl>     <dbl>
#> 1 <NA>      2/24/2003 0:00       10107 S10_1678                 30      95.7
#> 2 EMEA      5/7/2003 0:00        10121 S10_1678                 34      81.4
#> 3 EMEA      7/1/2003 0:00        10134 S10_1678                 41      94.7
#> 4 <NA>      8/25/2003 0:00       10145 S10_1678                 45      83.3
#> # … with 2,819 more rows
```

And here are the territories:

```
unique(sales$TERRITORY)
#> [1] NA      "EMEA" "APAC"  "Japan"
```

When I first started on this problem, I thought it was simple enough that I could just write the app without doing any other research:

```
ui <- fluidPage(
  selectInput("territory", "territory", choices = unique(sales$TERRITORY)),
  tableOutput("selected")
)
server <- function(input, output, session) {
  selected <- reactive(sales[sales$TERRITORY == input$territory, ])
  output$selected <- renderTable(head(selected(), 10))
}
```

I thought, *It's an eight-line app—what could possibly go wrong?* Well, when I opened the app up, I saw *a lot* of missing values, no matter what territory I selected. The code most likely to be the source of the problem was the reactive that selected the data to show: `sales[sales$TERRITORY == input$territory, ]`. So, I stopped the app and quickly verified that the subsetting worked the way I thought it did:

```
sales[sales$TERRITORY == "EMEA", ]
#> # A tibble: 2,481 x 6
#>    TERRITORY ORDERDATE     ORDERNUMBER PRODUCTCODE QUANTITYORDERED PRICEEACH
#>    <chr>     <chr>               <dbl> <chr>                 <dbl>     <dbl>
#> 1 <NA>      <NA>                   NA <NA>                    NA        NA
#> 2 EMEA      5/7/2003 0:00       10121 S10_1678                34      81.4
#> 3 EMEA      7/1/2003 0:00       10134 S10_1678                41      94.7
#> 4 <NA>      <NA>                   NA <NA>                    NA        NA
#> # … with 2,477 more rows
```

Oops! I'd forgotten that `TERRITORY` contained a bunch of missing values, which means that `sales$TERRITORY == "EMEA"` would contain a bunch of missing values:

```
head(sales$TERRITORY == "EMEA", 25)
#>  [1]    NA  TRUE  TRUE    NA    NA    NA  TRUE  TRUE    NA  TRUE FALSE    NA
#> [13]    NA    NA  TRUE    NA  TRUE  TRUE    NA    NA  TRUE FALSE  TRUE    NA
#> [25]  TRUE
```

These missing values become missing rows, and when I use them to subset the `sales` data frame with [, any missing values in input will be preserved in the output. There are lots of ways to resolve this problem, but I decided to use `subset()`[3] because it automatically removes missing values and reduces the number of times I need to type `sales`. I then double-checked this actually worked:

```
subset(sales, TERRITORY == "EMEA")
#> # A tibble: 1,407 x 6
#>    TERRITORY ORDERDATE     ORDERNUMBER PRODUCTCODE QUANTITYORDERED PRICEEACH
#>    <chr>     <chr>               <dbl> <chr>                 <dbl>     <dbl>
#> 1 EMEA      5/7/2003 0:00       10121 S10_1678                34      81.4
```

---

3 I'm using `subset()` so that my app doesn't require any other packages. In a bigger app, I'd probably prefer `dplyr::filter()` just because I'm a little more familiar with its behavior.

```
#> 2 EMEA      7/1/2003 0:00     10134 S10_1678              41       94.7
#> 3 EMEA      11/11/2003 0:00   10180 S10_1678              29       86.1
#> 4 EMEA      11/18/2003 0:00   10188 S10_1678              48       100
#> # … with 1,403 more rows
```

This fixed most of the problems, but I *still* had a problem when I selected NA in the territory drop-down: there were still no rows appearing. So again, I checked on the console:

```
subset(sales, TERRITORY == NA)
#> # A tibble: 0 x 6
#> # … with 6 variables: TERRITORY <chr>, ORDERDATE <chr>, ORDERNUMBER <dbl>,
#> #   PRODUCTCODE <chr>, QUANTITYORDERED <dbl>, PRICEEACH <dbl>
```

And then I remembered that of course this won't work because missing values are infectious:

```
head(sales$TERRITORY == NA, 25)
#>  [1] NA NA NA NA NA NA NA NA NA NA NA NA NA NA NA NA NA NA NA NA NA NA NA NA NA
```

There's another trick you can use to resolve this problem: switch from == to %in%:

```
head(sales$TERRITORY %in% NA, 25)
#>  [1]  TRUE FALSE FALSE  TRUE  TRUE  TRUE FALSE FALSE  TRUE FALSE FALSE  TRUE
#> [13]  TRUE  TRUE FALSE  TRUE FALSE FALSE  TRUE  TRUE FALSE FALSE FALSE  TRUE
#> [25] FALSE
subset(sales, TERRITORY %in% NA)
#> # A tibble: 1,074 x 6
#>   TERRITORY ORDERDATE        ORDERNUMBER PRODUCTCODE QUANTITYORDERED PRICEEACH
#>   <chr>     <chr>                  <dbl> <chr>                 <dbl>     <dbl>
#> 1 <NA>      2/24/2003 0:00         10107 S10_1678                30      95.7
#> 2 <NA>      8/25/2003 0:00         10145 S10_1678                45      83.3
#> 3 <NA>      10/10/2003 0:00        10159 S10_1678                49      100
#> 4 <NA>      10/28/2003 0:00        10168 S10_1678                36      96.7
#> # … with 1,070 more rows
```

So I updated the app and tried again. It still didn't work! When I selected "NA" in the drop-down, I didn't see any rows.

At this point, I figured I'd done everything I could on the console, and I needed to perform an experiment to figure out why the code inside of Shiny was working the way I expected. I guessed that the most likely source of the problem would be in the selected reactive, so I added a browser() statement there. This made it a two-line reactive, so I also needed to wrap it in {}:

```
server <- function(input, output, session) {
  selected <- reactive({
    browser()
    subset(sales, TERRITORY %in% input$territory)
  })
  output$selected <- renderTable(head(selected(), 10))
}
```

Now when my app ran, I was immediately dumped into an interactive console. My first step was to verify that I was in the problematic situation, so I ran `subset(sales, TERRITORY %in% input$territory)`. It returned an empty data frame, so I knew I was where I needed to be. If I hadn't seen the problem, I would have typed c to let the app continue running, then interacted with it some more in order to get it to the failing state.

I then checked that inputs to `subset()` were as I expected. I first double-checked that that the `sales` dataset looked OK. I didn't really expect it to be corrupted, since nothing in the app was touching it, but it's safest to carefully check every assumption that you're making. `sales` looked OK, so the problem must be in `TERRITORY %in% input$territory`. I started by inspecting `input$territory`, since `TERRITORY` is part of `sales`:

```
input$territory
#> [1] "NA"
```

I stared at this for a while, because it also looked OK. Then it occurred to me: I was expecting it to be NA, but it's actually "NA"! Now I could re-create the problem outside of the Shiny app:

```
subset(sales, TERRITORY %in% "NA")
#> # A tibble: 0 x 6
#> # … with 6 variables: TERRITORY <chr>, ORDERDATE <chr>, ORDERNUMBER <dbl>,
#> #   PRODUCTCODE <chr>, QUANTITYORDERED <dbl>, PRICEEACH <dbl>
```

Then I figured out a simple fix, applied it to my server, and reran the app:

```
server <- function(input, output, session) {
  selected <- reactive({
    if (input$territory == "NA") {
      subset(sales, is.na(TERRITORY))
    } else {
      subset(sales, TERRITORY == input$territory)
    }
  })
  output$selected <- renderTable(head(selected(), 10))
}
```

Hooray! The problem was fixed! But this felt pretty surprising to me: Shiny had silently converted an NA to an "NA", so I also filed a bug report (*https://oreil.ly/nZCg5*).

Several weeks later, I looked at this example again and started thinking about the different territories. We have Europe, Middle East, and Africa (EMEA) and Asia-Pacific (APAC). Where was North America? Then it dawned on me: the source data probably used the abbreviation NA, and R was reading it in as a missing value. So the real fix should happen during the data loading:

```
sales <- readr::read_csv("sales-dashboard/sales_data_sample.csv", na = "")
unique(sales$TERRITORY)
#> [1] "NA"    "EMEA"  "APAC"  "Japan"
```

That made life much simpler!

This is a common pattern when it comes to debugging: you often need to peel back multiple layers of the onion before you fully understand the source of the issue.

Debugging Reactivity

The hardest type of problem to debug is when your reactives fire in an unexpected order. At this point in the book, we have relatively few tools to recommend to help you debug this issue. In the next section, you'll learn how to create a minimal reprex, which is crucial for this type of problem, and later in the book, you'll learn more about the underlying theory and about tools like the reactive log (*https://github.com/rstudio/reactlog*). But for now, we'll focus on a classic technique that's useful here: "print" debugging.

The basic idea of print debugging is to call print() whenever you need to understand when a part of your code is evaluated and to show the values of important variables. We call this "print" debugging (because in most languages you'd use a print function), but in R, it makes more sense to use message():

- print() is designed for displaying vectors of data so it puts quotes around strings and starts the first line with [1].

- message() sends its result to "standard error," rather than "standard output." These are technical terms describing output streams, which you don't normally notice because they're both displayed in the same way when running interactively. But if your app is hosted elsewhere, then output sent to "standard error" will be recorded in the logs.

I also recommend coupling message() with glue::glue(), which makes it easy to interleave text and values in a message. If you haven't seen glue (*http://glue.tidyverse.org*) before, the basic idea is that anything wrapped inside {} will be evaluated and inserted into the output:

```
library(glue)
name <- "Hadley"
message(glue("Hello {name}"))
#> Hello Hadley
```

A final useful tool is str(), which prints the detailed structure of any object. This is particularly useful if you need to double-check you have the type of object that you expect.

Here's a toy app that shows off some of the basic ideas. Note how I use `message()` inside a `reactive()`: I have to perform the computation, send the message, and then return the previously computed value:

```r
ui <- fluidPage(
  sliderInput("x", "x", value = 1, min = 0, max = 10),
  sliderInput("y", "y", value = 2, min = 0, max = 10),
  sliderInput("z", "z", value = 3, min = 0, max = 10),
  textOutput("total")
)
server <- function(input, output, session) {
  observeEvent(input$x, {
    message(glue("Updating y from {input$y} to {input$x * 2}"))
    updateSliderInput(session, "y", value = input$x * 2)
  })

  total <- reactive({
    total <- input$x + input$y + input$z
    message(glue("New total is {total}"))
    total
  })

  output$total <- renderText({
    total()
  })
}
```

When I start the app, the console shows:

```
Updating y from 2 to 2
New total is 6
```

And if I drag the x slider to 3, I see:

```
Updating y from 2 to 6
New total is 8
New total is 12
```

Don't worry if you find the results a little surprising. You'll learn more about what's going on in Chapter 8 and "The Reactive Graph" on page 33.

Getting Help

If you're still stuck after trying these techniques, it's probably time to ask someone else. A great place to get help is the RStudio Community site (*https://oreil.ly/SXMCw*). This site is read by many Shiny users as well as the developers of the Shiny package itself. It's also a great place to visit if you want to improve your Shiny skills by helping others.

To get the most useful help as quickly as possible, you need to create a reprex, or **repr**oducible **ex**ample. The goal of a reprex is to provide the smallest possible snippet of R code that illustrates the problem and can easily be run on another computer. It's common courtesy (and in your own best interest) to create a reprex: if you want someone to help you, you should make it as easy as possible for them!

Making a reprex is polite because it captures the essential elements of the problem in a form that anyone else can run so that whoever attempts to help you can quickly see exactly what the problem is and can easily experiment with possible solutions.

Reprex Basics

A reprex is just some R code that works when you copy and paste it into an R session on another computer. Here's a simple Shiny app reprex:

```
library(shiny)
ui <- fluidPage(
  selectInput("n", "N", 1:10),
  plotOutput("plot")
)
server <- function(input, output, session) {
  output$plot <- renderPlot({
    n <- input$n * 2
    plot(head(cars, n))
  })
}
shinyApp(ui, server)
```

This code doesn't make any assumptions about the computer on which it's running (except that Shiny is installed!), so anyone can run this code and see the problem: the app throws an error saying non-numeric argument to binary operator.

Clearly illustrating the problem is the first step to getting help, and because anyone can reproduce the problem by just copying and pasting the code, they can easily explore your code and test possible solutions. (In this case, you need as.numeric(input$n) since selectInput() creates a string in input$n.)

Making a Reprex

The first step in making a reprex is to create a single self-contained file that contains everything needed to run your code. You should check that it works by starting a fresh R session and then running the code. Make sure you haven't forgotten to load any packages[4] that make your app work.

---

4 Regardless of how you normally load packages, I strongly recommend using multiple library() calls. This eliminates a source of potential confusion for people who might not be familiar with the tool that you're using.

Typically, the most challenging part of making your app work on someone else's computer is eliminating the use of data that's only stored on your computer. There are three useful patterns:

- Often the data you're using is not directly related to the problem, and you can instead use a built-in dataset like mtcars or iris.

- Other times, you might be able to write a little R code that creates a dataset that illustrates the problem:

  ```
  mydata <- data.frame(x = 1:5, y = c("a", "b", "c", "d", "e"))
  ```

- If both of those techniques fail, you can turn your data into code with dput(). For example, dput(mydata) generates the code that will re-create mydata:

  ```
  dput(mydata)
  #> structure(list(x = 1:5, y = c("a", "b", "c", "d", "e")),
  #> class = "data.frame", row.names = c(NA, -5L))
  ```

  Once you have that code, you can put this in your reprex to generate mydata:

  ```
  mydata <- structure(list(x = 1:5, y = structure(1:5, .Label = c("a", "b",
  "c", "d", "e"), class = "factor")), class = "data.frame", row.names =
  c(NA, -5L))
  ```

  Often, running dput() on your original data will generate a huge amount of code, so find a subset of your data that illustrates the problem. The smaller the dataset that you supply, the easier it will be for others to help you with your problem.

If reading data from a disk seems to be an irreducible part of the problem, a strategy of last resort is to provide a complete project containing both an *app.R* and the needed data files. The best way to provide this is as an RStudio project hosted on GitHub, but failing that, you can carefully make a zip file that can be run locally. Make sure that you use relative paths (i.e., read.csv("my-data.csv"), not read.csv("c:\ \my-user-name\\files\\my-data.csv")), so that your code still works when run on a different computer.

You should also consider the reader and spend some formatting your code so that it's easy to read. If you adopt the tidyverse style guide (*http://style.tidyverse.org*), you can automatically reformat your code using the styler (*http://styler.r-lib.org*) package; that quickly gets your code to a place that's easier to read.

Making a Minimal Reprex

Creating a reproducible example is a great first step because it allows someone else to precisely re-create your problem. However, the problematic code will often be buried among code that works just fine, so you can make the life of a helper much easier by trimming out the code that's OK.

Creating the smallest possible reprex is particularly important for Shiny apps, which are often complicated. You will get faster, higher-quality help if you can extract out the exact piece of the app that you're struggling with rather than forcing a potential helper to understand your entire app. As an added benefit, this process will often lead you to discover what the problem is so you don't have to wait for help from someone else!

Reducing a bunch of code to the essential problem is a skill, and you probably won't be very good at it at first. That's OK! Even the smallest reduction in code complexity helps the person helping you, and over time your reprex shrinking skills will improve.

If you don't know what part of your code is triggering the problem, a good way to find it is to remove sections of code from your application, piece by piece, until the problem goes away. If removing a particular piece of code makes the problem stop, it's likely that that code is related to the problem. Alternatively, sometimes it's simpler to start with a fresh, empty app and progressively build it up until you find the problem once more.

Once you've simplified your app to demonstrate the problem, it's worthwhile to take a final pass through and check:

- Is every input and output in UI related to the problem?
- Does your app have a complex layout that you can simplify to help focus on the problem at hand? Have you removed all UI customization that makes your app look good but isn't related to the problem?
- Are there any reactives in server() that you can now remove?
- If you've tried multiple ways to solve the problem, have you removed all the vestiges of the attempts that didn't work?
- Is every package that you load needed to illustrate the problem? Can you eliminate packages by replacing functions with dummy code?

This can be a lot of work, but the payoff is big: often you'll discover the solution while you make the reprex, and if not, it's much, much easier to get help.

Case Study

To illustrate the process of making a top-notch reprex, I'm going to use an example from Scott Novogoratz (*https://oreil.ly/FkqDU*) posted on RStudio Community (*https://oreil.ly/rJgWH*). The initial code was very close to being a reprex but wasn't quite reproducible because it forgot to load a pair of packages. As a starting point, I:

- Added missing library(lubridate) and library(xts).

- Split apart `ui` and `server` into separate objects.
- Reformatted the code with `styler::style_selection()`.

That yielded the following reprex:

```r
library(xts)
library(lubridate)
library(shiny)

ui <- fluidPage(
  uiOutput("interaction_slider"),
  verbatimTextOutput("breaks")
)
server <- function(input, output, session) {
  df <- data.frame(
    dateTime = c(
      "2019-08-20 16:00:00",
      "2019-08-20 16:00:01",
      "2019-08-20 16:00:02",
      "2019-08-20 16:00:03",
      "2019-08-20 16:00:04",
      "2019-08-20 16:00:05"
    ),
    var1 = c(9, 8, 11, 14, 16, 1),
    var2 = c(3, 4, 15, 12, 11, 19),
    var3 = c(2, 11, 9, 7, 14, 1)
  )

  timeSeries <- as.xts(df[, 2:4],
    order.by = strptime(df[, 1], format = "%Y-%m-%d %H:%M:%S")
  )
  print(paste(min(time(timeSeries)), is.POSIXt(min(time(timeSeries))), sep = " "))
  print(paste(max(time(timeSeries)), is.POSIXt(max(time(timeSeries))), sep = " "))

  output$interaction_slider <- renderUI({
    sliderInput(
      "slider",
      "Select Range:",
      min = min(time(timeSeries)),
      max = max(time(timeSeries)),
      value = c(min, max)
    )
  })

  brks <- reactive({
    req(input$slider)
    seq(input$slider[1], input$slider[2], length.out = 10)
  })

  output$breaks <- brks
}
shinyApp(ui, server)
```

If you run this reprex, you'll see the same problem in the initial post: an error stating "Type mismatch for min, max, and value. Each must be Date, POSIXt, or number." This is a solid reprex: I can easily run it on my computer, and it immediately illustrates the problem. However, it's a bit long, so it's not clear what's causing the problem.

To make this reprex simpler, we can carefully work through each line of code and see if it's important. While doing this, I discovered:

- Removing the two lines starting with `print()` didn't affect the error. Those two lines used `lubridate::is.POSIXt()`, which was the only use of lubridate, so once I removed them, I no longer needed to load lubridate.

- `df` is a data frame that's converted to an xts data frame called `timeSeries`. But the only way `timeSeries` is used is via `time(timeSeries)`, which returns a date-time. So I created a new variable `datetime` that contained some dummy date-time data. This still yielded the same error, so I removed `timeSeries` and `df`, and since that was the only place xts was used, I also removed `library(xts)`.

Together, those changes yielded a new `server()` that looked like this:

```r
datetime <- Sys.time() + (86400 * 0:10)

server <- function(input, output, session) {
  output$interaction_slider <- renderUI({
    sliderInput(
      "slider",
      "Select Range:",
      min   = min(datetime),
      max   = max(datetime),
      value = c(min, max)
    )
  })

  brks <- reactive({
    req(input$slider)
    seq(input$slider[1], input$slider[2], length.out = 10)
  })

  output$breaks <- brks
}
```

Next, I noticed that this example uses a relatively sophisticated Shiny technique where the UI is generated in the server function. But here `renderUI()` doesn't use any reactive inputs, so it should work the same way if moved out of the server function and into the UI.

This yielded a particularly nice result, because now the error occurs much earlier, before we even start the app:

```
ui <- fluidPage(
  sliderInput("slider",
    "Select Range:",
    min   = min(datetime),
    max   = max(datetime),
    value = c(min, max)
  ),
  verbatimTextOutput("breaks")
)
#> Error: Type mismatch for `min`, `max`, and `value`.
#> i All values must have same type: either numeric, Date, or POSIXt.
```

And now we can take the hint from the error message and look at each of the inputs we're feeding to min, max, and value to see where the problem is:

```
min(datetime)
#> [1] "2021-03-05 16:38:02 CST"
max(datetime)
#> [1] "2021-03-15 17:38:02 CDT"
c(min, max)
#> [[1]]
#> function (..., na.rm = FALSE)  .Primitive("min")
#>
#> [[2]]
#> function (..., na.rm = FALSE)  .Primitive("max")
```

Now the problem is obvious: we haven't assigned min and max variables, so we're accidentally passing the min() and max() functions into sliderInput(). One way to solve that problem is to use range() instead:

```
ui <- fluidPage(
  sliderInput("slider",
    "Select Range:",
    min   = min(datetime),
    max   = max(datetime),
    value = range(datetime)
  ),
  verbatimTextOutput("breaks")
)
```

This is a fairly typical outcome from creating a reprex: once you've simplified the problem to its key components, the solution becomes obvious. Creating a good reprex is an incredibly powerful debugging technique.

To simplify this reprex, I had to do a bunch of experimenting and reading up on functions that I wasn't very familiar with.[5] It's typically much easier to do this if it's your reprex, because you already understand the intent of the code. Still, you'll often need to do a bunch of experimentation to figure out where exactly the problem is

---

5 For example, I had no idea that is.POSIXt() was part of the lubridate package!

coming from. That can be frustrating and feel time-consuming, but it has a number of benefits:

- It enables you to create a description of the problem that is accessible to anyone who knows Shiny, not anyone who knows Shiny *and* the particular domain that you're working in.

- You will build up a better mental model of how your code works, which means that you're less likely to make the same or similar mistakes in the future.

- Over time, you'll get faster and faster at creating reprexes, and this will become one of your go-to techniques when debugging.

- Even if you don't create a perfect reprex, any work you can do to improve your reprex is less work for someone else to do. This is particularly important if you're trying to get help from package developers because they usually have many demands on their time.

When I try to help someone with their app on RStudio Community (*https://oreil.ly/ EHtZI*), creating a reprex is always the first thing I do. This isn't some make-work exercise I use to fob off people I don't want to help: it's exactly where I start!

Summary

This chapter has given you some useful workflows for developing apps, debugging problems, and getting help. These workflows might seem a little abstract and easy to dismiss because they're not concretely improving an individual app. But I think of workflow as one of my "secret" powers: one of the reasons that I've been able to accomplish so much is that I devote time to analyzing and improving my workflow. I highly encourage you to do the same!

The next chapter on layouts and themes is the first of a grab bag of useful techniques. There's no need to read in sequence; feel free to skip ahead to a chapter that you need for a current app.

## CHAPTER 6

Layout, Themes, HTML

Introduction

In this chapter you'll unlock some new tools for controlling the overall appearance of your app. We'll start by talking about page layouts (both single and "multiple") that let you organize your inputs and outputs. Then you'll learn about Bootstrap, the CSS toolkit that Shiny uses, and how to customize its overall visual appearance with themes. We'll finish with a brief discussion of what's going on under the hood so that if you know HTML and CSS, you can customize Shiny apps still further. As usual, we begin by loading shiny:

```
library(shiny)
```

Single-Page Layouts

In Chapter 2 you learned about the inputs and outputs that form the interactive components of the app. But I didn't talk about how to lay them out on the page, and instead I just used `fluidPage()` to slap them together as quickly as possible. While this is fine for learning Shiny, it doesn't create usable or visually appealing apps, so now it's time to learn some more layout functions.

Layout functions provide the high-level visual structure of an app. Layouts are created by a hierarchy of function calls, where the hierarchy in R matches the hierarchy in the generated HTML. This helps you understand layout code. For example, when you look at layout code like this:

```
fluidPage(
  titlePanel("Hello Shiny!"),
  sidebarLayout(
    sidebarPanel(
      sliderInput("obs", "Observations:", min = 0, max = 1000, value = 500)
```

```
  ),
    mainPanel(
      plotOutput("distPlot")
    )
  )
)
```

focus on the hierarchy of the function calls:

```
fluidPage(
  titlePanel(),
  sidebarLayout(
    sidebarPanel(),
    mainPanel()
  )
)
```

Even though you don't have these functions yet, you can guess what's going on by reading their names. You might imagine that this code will generate a classic app design: a title bar at top, followed by a sidebar (containing a slider) and main panel (containing a plot). The ability to easily see hierarchy through indentation is one of the reasons it's a good idea to use a consistent style.

In the remainder of this section, I'll discuss the functions that help you design single-page apps, then I'll move on to multipage apps in the next section. I also recommend checking out the Shiny Application layout guide (*https://oreil.ly/ZXl62*); it's a little dated but contains some useful gems.

Page Functions

The most important, but least interesting, layout function is `fluidPage()`, which you've seen in pretty much every example so far. But what's it doing, and what happens if you use it by itself? Figure 6-1 shows the results: it looks like a very boring app, but there's a lot going on behind the scenes, because `fluidPage()` sets up all the HTML, CSS, and JavaScript that Shiny needs.

In addition to `fluidPage()`, Shiny provides a couple of other page functions that can come in handy in more specialized situations: `fixedPage()` and `fillPage()`. `fixedPage()` works like `fluidPage()` but has a fixed maximum width, which stops your apps from becoming unreasonably wide on bigger screens. `fillPage()` fills the full height of the browser and is useful if you want to make a plot that occupies the whole screen. You can find the details in their documentation.

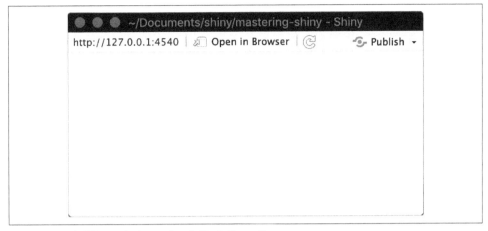

Figure 6-1. A UI consisting only of `fluidPage()`.

Page with Sidebar

To make more complex layouts, you'll need call layout functions inside of `fluid Page()`. For example, to make a two-column layout with inputs on the left and outputs on the right, you can use `sidebarLayout()` (along with its friends `titlePanel()`, `sidebarPanel()`, and `mainPanel()`). The following code shows the basic structure to generate Figure 6-2:

```
fluidPage(
  titlePanel(
    # app title/description
  ),
  sidebarLayout(
    sidebarPanel(
      # inputs
    ),
    mainPanel(
      # outputs
    )
  )
)
```

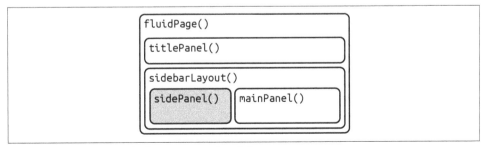

Figure 6-2. Structure of a basic app with sidebar.

To make it more realistic, let's add an input and output to create a very simple app that demonstrates the central limit theorem, as shown in Figure 6-3. If you run this app yourself, you can increase the number of samples to see the distribution become more normal:

```
ui <- fluidPage(
  titlePanel("Central limit theorem"),
  sidebarLayout(
    sidebarPanel(
      numericInput("m", "Number of samples:", 2, min = 1, max = 100)
    ),
    mainPanel(
      plotOutput("hist")
    )
  )
)
server <- function(input, output, session) {
  output$hist <- renderPlot({
    means <- replicate(1e4, mean(runif(input$m)))
    hist(means, breaks = 20)
  }, res = 96)
}
```

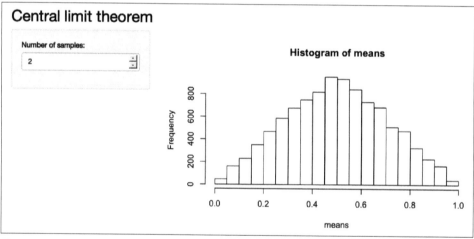

Figure 6-3. A common app design is to put controls in a sidebar and display results in the main panel.

Multirow

Under the hood, `sidebarLayout()` is built on top of a flexible multirow layout, which you can use directly to create more visually complex apps. Start with `fluidPage()`, then you create rows with `fluidRow()` and columns with `column()`. The following template generates the structure shown in Figure 6-4:

```
fluidPage(
  fluidRow(
    column(4,
      ...
    ),
    column(8,
      ...
    )
  ),
  fluidRow(
    column(6,
      ...
    ),
    column(6,
      ...
    )
  )
)
```

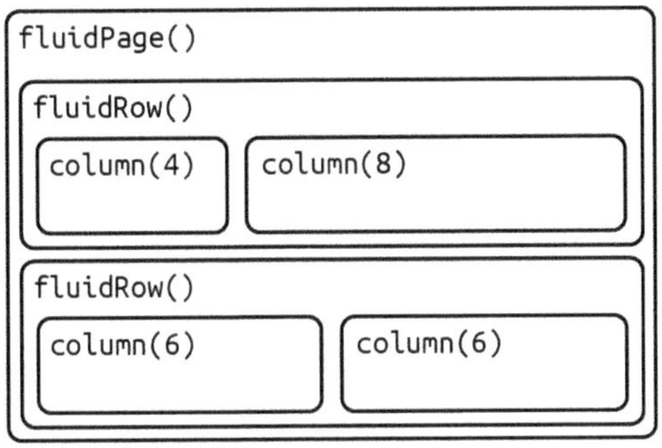

Figure 6-4. The structure underlying a simple multirow app.

Each row is made up of 12 columns, and the first argument to `column()` gives how many of those columns to occupy. A 12-column layout gives you substantial flexibility because you can easily create 2-, 3-, or 4-column layouts, or use narrow columns to create spacers. You can see an example of this layout in Figure 4-3.

If you'd like to learn more about designing using a grid system, I highly recommend the classic text on the subject, *Grid Systems in Graphic Design* by Josef Müller-Brockmann.

Exercises

1. Read the documentation of `sidebarLayout()` to determine the width (in columns) of the sidebar and the main panel. Can you re-create its appearance using `fluidRow()` and `column()`? What are you missing?

2. Modify the central limit theorem app to put the sidebar on the right instead of the left.

3. Create an app that contains two plots, each of which takes up half of the width. Put the controls in a full-width container below the plots.

Multipage Layouts

As your app grows in complexity, it might become impossible to fit everything on a single page. In this section you'll learn various uses of `tabPanel()` that create the illusion of multiple pages. This is an illusion because you'll still have a single app with a single underlying HTML file, but it's now broken into pieces, and only one piece is visible at a time.

Multipage apps pair particularly well with modules, which you'll learn about in Chapter 19. Modules allow you to partition the server function in the same way you partition the user interface, creating independent components that only interact through well-defined connections.

Tabsets

The simple way to break up a page into pieces is to use `tabsetPanel()` and its close friend `tabPanel()`. As you can see in the following code, `tabsetPanel()` creates a container for any number of `tabPanels()`, which can in turn contain any other HTML components. Figure 6-5 shows a simple example:

```
ui <- fluidPage(
  tabsetPanel(
    tabPanel("Import data",
      fileInput("file", "Data", buttonLabel = "Upload..."),
      textInput("delim", "Delimiter (leave blank to guess)", ""),
      numericInput("skip", "Rows to skip", 0, min = 0),
      numericInput("rows", "Rows to preview", 10, min = 1)
    ),
    tabPanel("Set parameters"),
    tabPanel("Visualise results")
  )
)
```

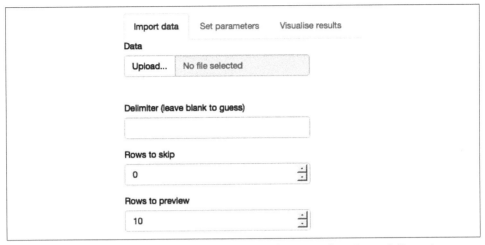

Figure 6-5. A tabsetPanel() allows the user to select a single tabPanel() to view.

If you want to know what tab a user has selected, you can provide the id argument to tabsetPanel() and it becomes an input. Figure 6-6 shows this in action:

```r
ui <- fluidPage(
  sidebarLayout(
    sidebarPanel(
      textOutput("panel")
    ),
    mainPanel(
      tabsetPanel(
        id = "tabset",
        tabPanel("panel 1", "one"),
        tabPanel("panel 2", "two"),
        tabPanel("panel 3", "three")
      )
    )
  )
)
server <- function(input, output, session) {
  output$panel <- renderText({
    paste("Current panel: ", input$tabset)
  })
}
```

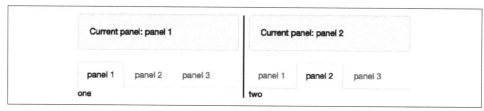

Figure 6-6. A tabset becomes an input when you use the `id` argument. This allows you to make your app behave differently depending on which tab is currently visible.

Note that `tabsetPanel()` can be used anywhere in your app; it's totally fine to nest tabsets inside of other components (including tabsets!) if needed.

Navlists and Navbars

Because tabs are displayed horizontally, there's a fundamental limit to how many tabs you can use, particularly if they have long titles. `navbarPage()` and `navbarMenu()` provide two alternative layouts that let you use more tabs with longer titles.

`navlistPanel()` is similar to `tabsetPanel()`, but instead of running the tab titles horizontally, it shows them vertically in a sidebar. It also allows you to add headings with plain strings, as shown in the following code, which generates Figure 6-7:

```
ui <- fluidPage(
  navlistPanel(
    id = "tabset",
    "Heading 1",
    tabPanel("panel 1", "Panel one contents"),
    "Heading 2",
    tabPanel("panel 2", "Panel two contents"),
    tabPanel("panel 3", "Panel three contents")
  )
)
```

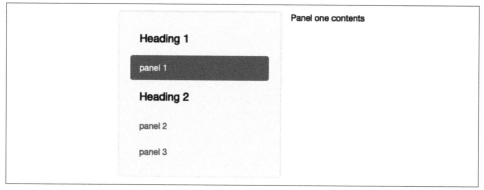

Figure 6-7. A `navlistPanel()` displays the tab titles vertically rather than horizontally.

Another approach is to use `navbarPage()`: it still runs the tab titles horizontally, but you can use `navbarMenu()` to add drop-down menus for an additional level of hierarchy. Figure 6-8 shows a simple example:

```
ui <- navbarPage(
  "Page title",
  tabPanel("panel 1", "one"),
  tabPanel("panel 2", "two"),
  tabPanel("panel 3", "three"),
  navbarMenu("subpanels",
    tabPanel("panel 4a", "four-a"),
    tabPanel("panel 4b", "four-b"),
    tabPanel("panel 4c", "four-c")
  )
)
```

Figure 6-8. A navbarPage() *sets up a horizontal nav bar at the top of the page.*

These layouts give you considerable ability to create rich and satisfying apps. To go further, you'll need to learn more about the underlying design system.

Bootstrap

To continue your app customization journey, you'll need to learn a little more about the Bootstrap (*https://getbootstrap.com*) framework used by Shiny. Bootstrap is a collection of HTML conventions, CSS styles, and JavaScript snippets bundled up into a convenient form. Bootstrap grew out of a framework originally developed for Twitter and over the last 10 years has grown to become one of the most popular CSS frameworks used on the web. Bootstrap is also popular in R—you've undoubtedly seen many documents produced by `rmarkdown::html_document()` and used many package websites made by pkgdown (*http://pkgdown.r-lib.org*), both of which also use Bootstrap.

As a Shiny developer, you don't need to think too much about Bootstrap, because Shiny functions automatically generate bootstrap-compatible HTML for you. But it's good to know that Bootstrap exists because then:

- You can use `bslib::bs_theme()` to customize the visual appearance of your code, as discussed in "Themes" on page 98.

- You can use the `class` argument to customize some layouts, inputs, and outputs using Bootstrap class names, as you saw in "Action Buttons" on page 20.

- You can make your own functions to generate Bootstrap components that Shiny doesn't provide, as explained in the Utility classes (*https://oreil.ly/ocDhF*) section of the bslib documentation.

It's also possible to use a completely different CSS framework. A number of existing R packages make this easy by wrapping popular alternatives to Bootstrap:

- shiny.semantic (*https://appsilon.github.io/shiny.semantic*), by Appsilon (*https://appsilon.com*), builds on top of Fomantic-UI (*https://fomantic-ui.com*).
- shinyMobile (*https://github.com/RinteRface/shinyMobile*), by RinteRface (*https://rinterface.com*), builds on top of Framework7 (*https://framework7.io*) and is specifically designed for mobile apps.
- shinymaterial (*https://ericrayanderson.github.io/shinymaterial*), by Eric Anderson (*https://github.com/ericrayanderson*), is built on top of Google's Material design (*https://material.io/design*) framework.
- shinydashboard (*https://rstudio.github.io/shinydashboard*), also by RStudio, provides a layout system designed to create dashboards.

You can find a fuller, and actively maintained, list on GitHub (*https://github.com/nanxstats/awesome-shiny-extensions*).

Themes

Bootstrap is so ubiquitous within the R community that it's easy to get style fatigue: after a while, every Shiny app and Rmd start to look the same. The solution is theming with the bslib (*https://rstudio.github.io/bslib*) package. bslib is a relatively new package that allows you to override many Bootstrap defaults in order to create an appearance that is uniquely yours. As I write this, bslib is mostly applicable only to Shiny, but work is afoot to bring its enhanced theming power to RMarkdown, pkgdown, and more.

If you're producing apps for your company, I highly recommend investing a little time in theming—theming your app to match your corporate style guide is an easy way to make yourself look good.

Getting Started

Create a theme with `bslib::bs_theme()`, then apply it to an app with the `theme` argument of the page layout function:

```
fluidPage(
  theme = bslib::bs_theme(...)
)
```

If not specified, Shiny will use the classic Bootstrap v3 theme that it has used basically since it was created. By default, `bslib::bs_theme()` will use Bootstrap v4. Using Bootstrap v4 instead of v3 will not cause problems if you only use built-in components. There is a possibility that it might cause problems if you've used custom HTML, so you can force it to stay with v3 with `version = 3`.

Shiny Themes

The easiest way to change the overall look of your app is to pick a premade "bootswatch" theme (*https://bootswatch.com*) using the `bootswatch` argument to `bslib::bs_theme()`. Figure 6-9 shows the results of the following code, switching `"darkly"` out for other themes:

```r
ui <- fluidPage(
  theme = bslib::bs_theme(bootswatch = "darkly"),
  sidebarLayout(
    sidebarPanel(
      textInput("txt", "Text input:", "text here"),
      sliderInput("slider", "Slider input:", 1, 100, 30)
    ),
    mainPanel(
      h1(paste0("Theme: darkly")),
      h2("Header 2"),
      p("Some text")
    )
  )
)
```

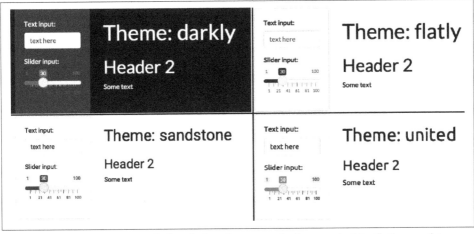

Figure 6-9. The same app styled with four bootswatch themes: darkly, flatly, sandstone, and united.

Alternatively, you can construct your own theme using the other arguments to bs_theme() like bg (background color), fg (foreground color), and base_font:[1]

```
theme <- bslib::bs_theme(
  bg = "#0b3d91",
  fg = "white",
  base_font = "Source Sans Pro"
)
```

An easy way to preview and customize your theme is to use bslib::bs_theme_pre view(theme). This will open a Shiny app that shows what the theme looks like when many standard controls are applied and also provides you with interactive controls for customizing the most important parameters.

Plot Themes

If you've heavily customized the style of your app, you may want to also customize your plots to match. Luckily, this is really easy thanks to the thematic (*https://rstu dio.github.io/thematic*) package, which automatically themes ggplot2, lattice, and base plots. Just call thematic_shiny() in your server function. This will automatically determine all of the settings from your app theme, as in Figure 6-10:

```
library(ggplot2)

ui <- fluidPage(
  theme = bslib::bs_theme(bootswatch = "darkly"),
  titlePanel("A themed plot"),
  plotOutput("plot"),
)

server <- function(input, output, session) {
  thematic::thematic_shiny()

  output$plot <- renderPlot({
    ggplot(mtcars, aes(wt, mpg)) +
      geom_point() +
      geom_smooth()
  }, res = 96)
}
```

---

1 Fonts are a little trickier than colors because you have to make sure the app viewer also has the font. Make sure to read the bs_theme() docs for all the details.

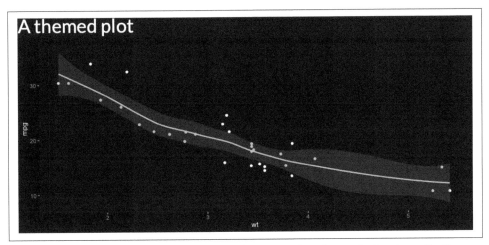

Figure 6-10. `thematic::thematic_shiny()` ensures that the ggplot2 automatically matches the app theme.

Exercises

1. Use `bslib::bs_theme_preview()` to make the ugliest theme you can.

Under the Hood

Shiny is designed so that, as an R user, you don't need to learn about the details of HTML. However, if you know some HTML and CSS, it's possible to customize Shiny still further. Unfortunately, teaching HTML and CSS is beyond the scope of this book, but the HTML (*https://oreil.ly/ig3dW*) and CSS basics (*https://oreil.ly/4Utd6*) tutorials by MDN are good places to start.

The most important thing to know is that there's no magic behind all the input, output, and layout functions: they just generate HTML.[2] You can see that HTML by executing UI functions directly in the console:

```
fluidPage(
  textInput("name", "What's your name?")
)

<div class="container-fluid">
  <div class="form-group shiny-input-container">
    <label for="name">What's your name?</label>
    <input id="name" type="text" class="form-control" value=""/>
```

---

2 The magic that connects inputs and outputs to R happens elsewhere (via JavaScript), but that's well beyond the scope of this book.

```
    </div>
  </div>
```

Note that this is the contents of the <body> tag; other parts of Shiny take care of generating the <head>. If you want to include additional CSS or JavaScript dependencies, you'll need to learn `htmltools::htmlDependency()`. Two good places to start are the R-hub blog post "JavaScript for the R Package Developer" (*https://oreil.ly/vUVAj*) and Chapter 4 of *Outstanding User Interfaces with Shiny* (*https://oreil.ly/eOYCN*).

It's possible to add your own HTML to the `ui`. One way to do so is by including literal HTML with the `HTML()` function. In the next example, I use the "raw character constant,"[3] `r"()"`, to make it easier to include quotes in the string:

```
ui <- fluidPage(
  HTML(r"(
    <h1>This is a heading</h1>
    <p class="my-class">This is some text!</p>
    <ul>
      <li>First bullet</li>
      <li>Second bullet</li>
    </ul>
  )")
)
```

If you're an HTML/CSS expert, you might be interested to know that you can skip `fluidPage()` altogether and supply raw HTML. See "Build Your Entire UI with HTML" (*https://oreil.ly/7UCaY*) for more details.

Alternatively, you can use the HTML helper that Shiny provides. There are regular functions for the most important elements like `h1()` and `p()`, and all others can be accessed via the tags helper. Named arguments become attributes, and unnamed arguments become children, so we can re-create the preceding HTML as:

```
ui <- fluidPage(
  h1("This is a heading"),
  p("This is some text", class = "my-class"),
  tags$ul(
    tags$li("First bullet"),
    tags$li("Second bullet")
  )
)
```

---

3 Introduced in R 4.0.0.

One advantage of generating HTML with code is that you can interweave existing Shiny components into a custom structure. For example, the following code makes a paragraph of text containing two outputs, one that is bold:

```
tags$p(
  "You made ",
  tags$b("$", textOutput("amount", inline = TRUE)),
  " in the last ",
  textOutput("days", inline = TRUE),
  " days "
)
```

Note the use of `inline = TRUE`; the `textOutput()` default is to produce a complete paragraph.

If you want to learn more about using HTML, CSS, and JavaScript to make compelling user interfaces, I highly recommend David Granjon's *Outstanding User Interfaces with Shiny* (*https://oreil.ly/q7aKy*).

Summary

This chapter has given you the tools you need to make complex and attractive Shiny apps. You've learned the Shiny functions that allow you to layout single and multipage apps (like `fluidPage()` and `tabsetPanel()`) and how to customize the overall visual appearance with themes. You've also learned a little bit about what's going on under the hood: you know that Shiny uses Bootstrap and that the input and output functions just return HTML, which you can also create yourself.

In the next chapter you'll learn more about another important visual component of your app: graphics.

## CHAPTER 7

Graphics

We talked briefly about `renderPlot()` in Chapter 2; it's a powerful tool for displaying graphics in your app. This chapter will show you how to use it to its full extent to create interactive plots, plots that respond to mouse events. You'll also learn a couple of other useful techniques, including making plots with dynamic width and height and displaying images with the `renderImage()`.

In this chapter, we'll need ggplot2 as well as shiny, since that's what I'll use for the majority of the graphics:

```
library(shiny)
library(ggplot2)
```

Interactivity

One of the coolest things about `plotOutput()` is that as well as being an output that displays plots, it can also be an input that responds to pointer events. That allows you to create interactive graphics where the user interacts directly with the data on the plot. Interactive graphics are a powerful tool, with a wide range of applications. I don't have space to show you all the possibilities, so here I'll focus on the basics, then point you toward resources to learn more.

Basics

A plot can respond to four different mouse[1] events: `click`, `dblclick` (double-click), `hover` (when the mouse stays in the same place for a little while), and `brush` (a

---

1 When I wrote this chapter, Shiny didn't support touch events, which means that plot interactivity won't work on mobile devices. Hopefully it will support these events by the time you read this.

rectangular selection tool). To turn these events into Shiny inputs, you supply a string to the corresponding plotOutput() argument—for example, plotOutput("plot", click = "plot_click"). This creates an input$plot_click that you can use to handle mouse clicks on the plot.

Here's a very simple example of handling a mouse click. We register the plot_click input and then use that to update an output with the coordinates of the mouse click. Figure 7-1 shows the results:

```
ui <- fluidPage(
  plotOutput("plot", click = "plot_click"),
  verbatimTextOutput("info")
)

server <- function(input, output) {
  output$plot <- renderPlot({
    plot(mtcars$wt, mtcars$mpg)
  }, res = 96)

  output$info <- renderPrint({
    req(input$plot_click)
    x <- round(input$plot_click$x, 2)
    y <- round(input$plot_click$y, 2)
    cat("[", x, ", ", y, "]", sep = "")
  })
}
```

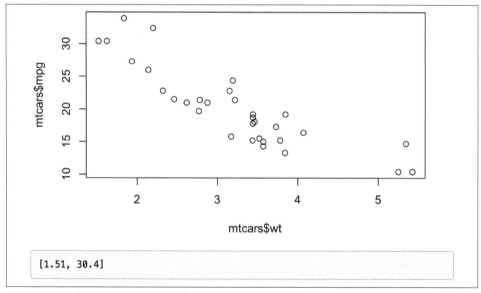

Figure 7-1. Clicking on the top-left point updates the printed coordinates. See live at https://hadley.shinyapps.io/ms-click.

(Note the use of `req()` to make sure the app doesn't do anything before the first click and that the coordinates are in terms of the underlying wt and mpg variables.)

The following sections describe the events in more detail. We'll start with the click events, then briefly discuss the closely related dblclick and hover. Then you'll learn about the brush event, which provides a rectangular "brush" defined by its four sides (xmin, xmax, ymin, and ymax). I'll then give a couple of examples of updating the plot with the results of the action, and then discuss some of the limitations of interactive graphics in Shiny.

Clicking

The point events return a relatively rich list containing a lot of information. The most important components are x and y, which give the location of the event in data coordinates. But I'm not going to talk about this data structure since you'll only need it in relatively rare situations. (If you do want the details, use this app (*https://oreil.ly/KkhOH*) in the Shiny gallery.) Instead, you'll use the nearPoints() helper, which returns a data frame containing rows near[2] the click, taking care of a bunch of fiddly details.

Here's a simple example of nearPoints() in action, showing a table of data about the points near the event. Figure 7-2 shows a screenshot of the app:

```
ui <- fluidPage(
  plotOutput("plot", click = "plot_click"),
  tableOutput("data")
)
server <- function(input, output, session) {
  output$plot <- renderPlot({
    plot(mtcars$wt, mtcars$mpg)
  }, res = 96)

  output$data <- renderTable({
    nearPoints(mtcars, input$plot_click, xvar = "wt", yvar = "mpg")
  })
}
```

---

2 Note that it's not called nearestPoints(); it won't return anything if you don't click near an existing data point.

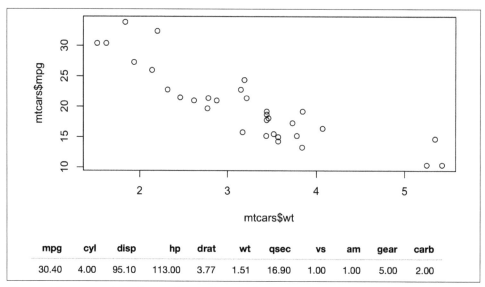

mpg	cyl	disp	hp	drat	wt	qsec	vs	am	gear	carb
30.40	4.00	95.10	113.00	3.77	1.51	16.90	1.00	1.00	5.00	2.00

Figure 7-2. nearPoints() translates plot coordinates to data rows, making it easy to show the underlying data for a point you clicked on. See live at https:// hadley.shinyapps.io/ms-nearPoints.

Here we give nearPoints() four arguments: the data frame that underlies the plot, the input event, and the names of the variables on the axes. If you use ggplot2, you only need to provide the first two arguments since xvar and yvar can be automatically imputed from the plot data structure. For that reason, I'll use ggplot2 throughout the rest of the chapter. Here's that previous example reimplemented with ggplot2:

```
ui <- fluidPage(
  plotOutput("plot", click = "plot_click"),
  tableOutput("data")
)
server <- function(input, output, session) {
  output$plot <- renderPlot({
    ggplot(mtcars, aes(wt, mpg)) + geom_point()
  }, res = 96)

  output$data <- renderTable({
    req(input$plot_click)
    nearPoints(mtcars, input$plot_click)
  })
}
```

You might wonder exactly what `nearPoints()` returns. This is a good place to use `browser()`, which we discussed in "The Interactive Debugger" on page 74:

```
...
  output$data <- renderTable({
    req(input$plot_click)
    browser()
    nearPoints(mtcars, input$plot_click)
  })
...
```

Now after I start the app and click on a point, I'm dropped into the interactive debugger, where I can run `nearPoints()` and see what it returns:

```
nearPoints(mtcars, input$plot_click)
#>              mpg cyl disp hp drat   wt  qsec vs am gear carb
#> Datsun 710 22.8   4  108 93 3.85 2.32 18.61  1  1    4    1
```

Another way to use `nearPoints()` is with `allRows = TRUE` and `addDist = TRUE`. That will return the original data frame with two new columns:

- `dist_` gives the distance between the row and the event (in pixels).
- `selected_` says whether or not it's near the click event (i.e., whether or not it's a row that would be returned when `allRows = FALSE`).

We'll see an example of that a little later.

Other Point Events

The same approach works equally well with `click`, `dblclick`, and `hover`: just change the name of the argument. If needed, you can get additional control over the events by supplying `clickOpts()`, `dblclickOpts()`, or `hoverOpts()` instead of a string giving the input ID. These are rarely needed, so I won't discuss them here; see the documentation for details.

You can use multiple interaction types on one plot. Just make sure to explain to the user what they can do: one downside of using mouse events to interact with an app is that they're not immediately discoverable.[3]

Brushing

Another way of selecting points on a plot is to use a *brush*, a rectangular selection defined by four edges. In Shiny, using a brush is straightforward once you've

---

3 As a general rule, adding explanatory text suggests that your interface is too complex, so it's best avoided, where possible. This is the key idea behind *affordances*, the idea that an object should suggest naturally how to interact with it, as introduced by Don Norman in *The Design of Everyday Things*.

mastered `click` and `nearPoints()`: you just switch to `brush` argument and the `brush` `edPoints()` helper.

Here's another simple example that shows which points have been selected by the brush. Figure 7-3 shows the results:

```
ui <- fluidPage(
  plotOutput("plot", brush = "plot_brush"),
  tableOutput("data")
)
server <- function(input, output, session) {
  output$plot <- renderPlot({
    ggplot(mtcars, aes(wt, mpg)) + geom_point()
  }, res = 96)

  output$data <- renderTable({
    brushedPoints(mtcars, input$plot_brush)
  })
}
```

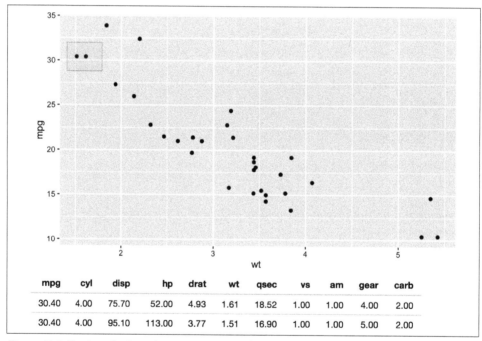

mpg	cyl	disp	hp	drat	wt	qsec	vs	am	gear	carb
30.40	4.00	75.70	52.00	4.93	1.61	18.52	1.00	1.00	4.00	2.00
30.40	4.00	95.10	113.00	3.77	1.51	16.90	1.00	1.00	5.00	2.00

Figure 7-3. Setting the `brush` argument provides the user with a draggable "brush." In this app, the points beneath the brush are shown in a table. See live at https:// hadley.shinyapps.io/ms-brushedPoints.

Use brushOpts() to control the color (fill and stroke), or restrict brushing to a single dimension with direction = "x" or "y" (useful, for example, for brushing time series).

Modifying the Plot

So far we've displayed the results of the interaction in another output. But the true beauty of interactivity comes when you display the changes in the same plot you're interacting with. Unfortunately, this requires an advanced reactivity technique that you have not yet learned about: reactiveVal(). We'll come back to reactiveVal() in Chapter 16, but I wanted to show it here because it's such a useful technique. You'll probably need to reread this section after you've read that chapter, but hopefully, even without all the theory, you'll get a sense of the potential applications.

As you might guess from the name, reactiveVal() is rather similar to reactive(). You create a reactive value by calling reactiveVal() with its initial value, and retrieve that value in the same way as a reactive:

```
val <- reactiveVal(10)
val()
#> [1] 10
```

The big difference is that you can also update a reactive value, and all reactive consumers that refer to it will recompute. A reactive value uses a special syntax for updating—you call it like a function, with the first argument being the new value:

```
val(20)
val()
#> [1] 20
```

That means updating a reactive value using its current value looks something like this:

```
val(val() + 1)
val()
#> [1] 21
```

Unfortunately, if you actually try to run this code in the console, you'll get an error because it has to be run in a reactive environment. That makes experimentation and debugging more challenging because you'll need to use browser() or similar to pause execution within the call to shinyApp(). This is one of the challenges we'll come back to later in Chapter 16.

For now, let's put the challenges of learning reactiveVal() aside and show you why you might bother. Imagine that you want to visualize the distance between a click and the points on the plot. In the following app, we start by creating a reactive value to store those distances, initializing it with a constant that will be used before we click anything. Then we use observeEvent() to update the reactive value when the mouse is clicked and create a ggplot that visualizes the distance with point size. All up, this looks something like the following code and results in Figure 7-4:

```
set.seed(1014)
df <- data.frame(x = rnorm(100), y = rnorm(100))

ui <- fluidPage(
  plotOutput("plot", click = "plot_click", )
)
server <- function(input, output, session) {
  dist <- reactiveVal(rep(1, nrow(df)))
  observeEvent(input$plot_click,
    dist(nearPoints(df, input$plot_click, allRows = TRUE, addDist = TRUE)$dist_)
  )

  output$plot <- renderPlot({
    df$dist <- dist()
    ggplot(df, aes(x, y, size = dist)) +
      geom_point() +
      scale_size_area(limits = c(0, 1000), max_size = 10, guide = NULL)
  }, res = 96)
}
```

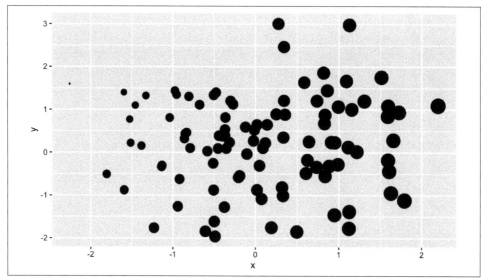

Figure 7-4. This app uses a `reactiveVal()` to store the distance to the point that was last clicked, which is then mapped to point size. Here I show the results of clicking on a point on the far left. See live at https://hadley.shinyapps.io/ms-modifying-size.

There are two important ggplot2 techniques to note here:

- I add the distances to the data frame before plotting. I think it's good practice to put related variables together in a data frame before visualizing it.

- I set the `limits` to `scale_size_area()` to ensure that sizes are comparable across clicks. To find the correct range, I did a little interactive experimentation, but you can work out the exact details if needed (see the exercises at the end of the chapter).

Here's a more complicated idea. I want to use a brush to progressively add points to a selection. Here I display the selection using different colors, but you could imagine many other applications. To make this work, I initialize the `reactiveVal()` to a vector of `FALSE`s, then use `brushedPoints()` and | to add any points under the brush to the selection. To give the user some way to start afresh, I make double-clicking reset the selection. Figure 7-5 shows a couple of screenshots from the running app:

```
ui <- fluidPage(
  plotOutput("plot", brush = "plot_brush", dblclick = "plot_reset")
)
server <- function(input, output, session) {
  selected <- reactiveVal(rep(FALSE, nrow(mtcars)))

  observeEvent(input$plot_brush, {
    brushed <- brushedPoints(mtcars, input$plot_brush, allRows = TRUE)$selected_
    selected(brushed | selected())
  })
  observeEvent(input$plot_reset, {
    selected(rep(FALSE, nrow(mtcars)))
  })

  output$plot <- renderPlot({
    mtcars$sel <- selected()
    ggplot(mtcars, aes(wt, mpg)) +
      geom_point(aes(colour = sel)) +
      scale_colour_discrete(limits = c("TRUE", "FALSE"))
  }, res = 96)
}
```

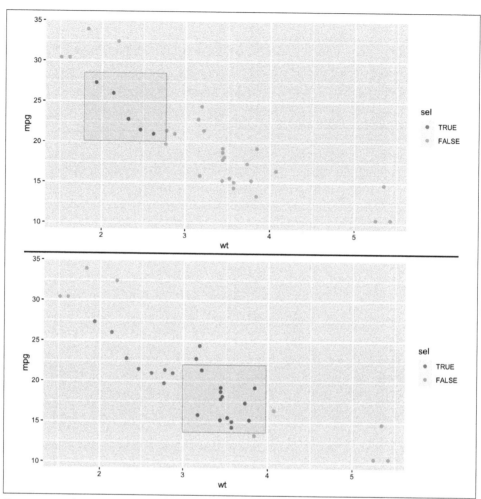

Figure 7-5. This app makes the brush "persistent" so that dragging it adds to the current selection.

Again, I set the limits of the scale to ensure that the legend (and colors) don't change after the first click.

Interactivity Limitations

Before we move on, it's important to understand the basic data flow in interactive plots in order to understand their limitations. The basic flow is something like this:

1. JavaScript captures the mouse event.
2. Shiny sends the mouse event data back to R, telling the app that the input is now out of date.
3. All the downstream reactive consumers are recomputed.
4. `plotOutput()` generates a new PNG and sends it to the browser.

For local apps, the bottleneck tends to be the time taken to draw the plot. Depending on how complex the plot is, this may take a significant fraction of a second. But for hosted apps, you also have to take into account the time needed to transmit the event from the browser to R and then the rendered plot back from R to the browser.

In general, this means that it's not possible to create Shiny apps where action and response is perceived as instantaneous (i.e., the plot appears to update simultaneously with your action upon it). If you need that level of speed, you'll have to perform more computation in JavaScript. One way to do this is to use an R package that wraps a JavaScript graphics library. Right now, as I write this book, I think you'll get the best experience with the plotly package, as documented in the book *Interactive Web-Based Data Visualization with R, Plotly, and Shiny* (*https://plotly-r.com*), by Carson Sievert.

Dynamic Height and Width

The rest of this chapter is less exciting than interactive graphics but contains material that's important to cover somewhere.

First of all, it's possible to make the plot size reactive so the width and height changes in response to user actions. To do this, supply zero-argument functions to the `width` and `height` arguments of `renderPlot()`—these now must be defined in the server, not the UI, since they can change. These functions should have no argument and return the desired size in pixels. They are evaluated in a reactive environment so that you can make the size of your plot dynamic.

The following app illustrates the basic idea. It provides two sliders that directly control the size of the plot. A couple of sample screenshots are shown in Figure 7-6. Note that when you resize the plot, the data stays the same: you don't get new random numbers:

```
ui <- fluidPage(
  sliderInput("height", "height", min = 100, max = 500, value = 250),
  sliderInput("width", "width", min = 100, max = 500, value = 250),
  plotOutput("plot", width = 250, height = 250)
```

```
)
server <- function(input, output, session) {
  output$plot <- renderPlot(
    width = function() input$width,
    height = function() input$height,
    res = 96,
    {
      plot(rnorm(20), rnorm(20))
    }
  )
}
```

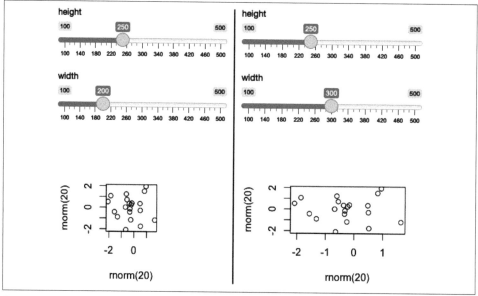

Figure 7-6. You can make the plot size dynamic so that it responds to user actions. This figure shows off the effect of changing the width. See live at https://hadley.shinyapps.io/ ms-resize.

In real apps, you'll use more complicated expressions in the width and height functions. For example, if you're using a faceted plot in ggplot2, you might use it to increase the size of the plot to keep the individual facet sizes roughly the same.[4]

Images

You can use renderImage() if you want to display existing images (not plots). For example, you might have a directory of photographs that you want shown to the user.

---

4 Unfortunately, there's no easy way to keep them exactly the same because it's currently not possible to find out the size of the fixed elements around the borders of the plot.

The following app illustrates the basics of renderImage() by showing cute puppy photos (Figure 7-7). The photos come from Unsplash (*https://unsplash.com*), my favorite source of royalty-free stock photographs:

```r
puppies <- tibble::tribble(
  ~breed, ~ id, ~author,
  "corgi", "eoqnr8ikwFE","alvannee",
  "labrador", "KCdYn0xu2fU", "shaneguymon",
  "spaniel", "TzjMd7i5WQI", "_redo_"
)

ui <- fluidPage(
  selectInput("id", "Pick a breed", choices = setNames(puppies$id, puppies$breed)),
  htmlOutput("source"),
  imageOutput("photo")
)
server <- function(input, output, session) {
  output$photo <- renderImage({
    list(
      src = file.path("puppy-photos", paste0(input$id, ".jpg")),
      contentType = "image/jpeg",
      width = 500,
      height = 650
    )
  }, deleteFile = FALSE)

  output$source <- renderUI({
    info <- puppies[puppies$id == input$id, , drop = FALSE]
    HTML(glue::glue("<p>
      <a href='https://unsplash.com/photos/{info$id}'>original</a> by
      <a href='https://unsplash.com/@{info$author}'>{info$author}</a>
    </p>"))
  })
}
```

renderImage() needs to return a list. The only crucial argument is src, a local path to the image file. You can additionally supply:

- A contentType, which defines the MIME type of the image. If not provided, Shiny will guess from the file extension, so you only need to supply this if your images don't have extensions.

- The width and height of the image, if known.

- Any other arguments, like class or alt, will be added as attributes to the <img> tag in the HTML.

Figure 7-7. An app that displays cute pictures of puppies using renderImage(). *See live at https://hadley.shinyapps.io/ms-puppies.*

You *must* also supply the deleteFile argument. Unfortunately, renderImage() was originally designed to work with temporary files, so it automatically deleted images after rendering them. This was obviously very dangerous, so the behavior changed in Shiny 1.5.0. Now Shiny no longer deletes the images but instead forces you to explicitly choose which behavior you want.

You can learn more about renderImage() and see other ways that you might use it on the Shiny website (*https://oreil.ly/zgzNm*).

Summary

Visualizations are a tremendously powerful way to communicate data, and this chapter has given you a few advanced techniques to empower your Shiny apps. Next, you'll learn techniques to provide feedback to your users about what's going on in your app, which is particularly important for actions that take a nontrivial amount of time.

## CHAPTER 8

User Feedback

You can often make your app more usable by giving the user more insight into what is happening. This might take the form of better messages when inputs don't make sense, or progress bars for operations that take a long time. Some feedback occurs naturally through outputs, which you already know how to use, but you'll often need something else. The goal of this chapter is to show you some of your other options.

We'll start with techniques for *validation*, informing the user when an input (or combination of inputs) is in an invalid state. We'll then continue on to *notification*, sending general messages to the user, and *progress bars*, which give details for time-consuming operations made up of many small steps. We'll finish up by discussing dangerous actions and how you give your users peace of mind with *confirmation* dialogs or the ability to *undo* an action.

In this chapter we'll use shinyFeedback (*https://oreil.ly/luGeN*), by Andy Merlino, and waiter (*http://waiter.john-coene.com*), by John Coene. You should also keep your eyes open for shinyvalidate (*https://oreil.ly/XADgf*), a package by Joe Cheng, which is currently under development. Let's begin by loading shiny:

```
library(shiny)
```

Validation

The first and most important feedback you can give to the user is that they've given you bad input. This is analogous to writing good functions in R: user-friendly functions give clear error messages describing what the expected input is and how you have violated those expectations. Thinking through how the user might misuse your app allows you to provide informative messages in the UI rather than allowing errors to trickle through into the R code and generate uninformative errors.

Validating Input

A great way to give additional feedback to the user is via the shinyFeedback (*https:// oreil.ly/luGeN*) package. Using it is a two-step process. First, you add `useShinyFeed back()` to the `ui`. This sets up the needed HTML and JavaScript for attractive error message display:

```
ui <- fluidPage(
  shinyFeedback::useShinyFeedback(),
  numericInput("n", "n", value = 10),
  textOutput("half")
)
```

Then in your `server()` function, you call one of the feedback functions: `feedback()`, `feedbackWarning()`, `feedbackDanger()`, and `feedbackSuccess()`. They all have three key arguments:

`inputId`
> The ID of the input where the feedback should be placed.

`show`
> A logical value determining whether or not to show the feedback.

`text`
> The text to display.

They also have `color` and `icon` arguments that you can use to further customize the appearance. See the documentation for more details.

Let's see how this comes together in a real example, pretending that we only want to allow even numbers:

```
server <- function(input, output, session) {
  half <- reactive({
    even <- input$n %% 2 == 0
    shinyFeedback::feedbackWarning("n", !even, "Please select an even number")
    input$n / 2
  })

  output$half <- renderText(half())
}
```

Figure 8-1 shows the results.

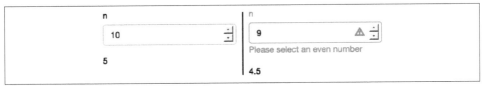

Figure 8-1. feedbackWarning() displays a warning for invalid inputs. The app on the left shows a valid input, and the app on the right shows an invalid (odd) input with warning message. See live at https://hadley.shinyapps.io/ms-feedback.

Notice that the error message is displayed but the output is still updated. Typically you don't want that because invalid inputs are likely to cause uninformative R errors that you don't want to show to the user. To stop inputs from triggering reactive changes, you need a new tool: req(), short for "required." It looks like this:

```
server <- function(input, output, session) {
  half <- reactive({
    even <- input$n %% 2 == 0
    shinyFeedback::feedbackWarning("n", !even, "Please select an even number")
    req(even)
    input$n / 2
  })

  output$half <- renderText(half())
}
```

When the input to req() is not true, it sends a special signal to tell Shiny that the reactive does not have all the inputs that it requires, so it should be "paused." We'll take a brief digression to talk about this before we come back to using it in concert with validate().

Canceling Execution with req()

It's easiest to understand req() by starting outside of validation. You may have noticed that when you start an app, the complete reactive graph is computed even before the user does anything. This works well when you can choose meaningful default values for your inputs. But that's not always possible, and sometimes you want to wait until the user actually does something. This tends to crop up with three controls:

- In textInput(), when you have used value = "" and you don't want to do anything until the user types something.
- In selectInput(), when you have provided an empty choice, "", and you don't want to do anything until the user makes a selection.

- In `fileInput()`, which has an empty result before the user has uploaded anything. We'll come back to this in Figure 9-1.

We need some way to "pause" reactives so that nothing happens until some condition is true. That's the job of `req()`, which checks for required values before allowing a reactive producer to continue.

For example, consider the following app, which will generate a greeting in English or Maori. If you run this app, you'll see an error, as in Figure 8-2, because there's no entry in the `greetings` vector that corresponds to the default choice of `""`:

```
ui <- fluidPage(
  selectInput("language", "Language", choices = c("", "English", "Maori")),
  textInput("name", "Name"),
  textOutput("greeting")
)

server <- function(input, output, session) {
  greetings <- c(
    English = "Hello",
    Maori = "Ki ora"
  )
  output$greeting <- renderText({
    paste0(greetings[[input$language]], " ", input$name, "!")
  })
}
```

Figure 8-2. The app displays an uninformative error when it is loaded because a language hasn't been selected yet.

We can fix this problem by using `req()`. Now nothing will be displayed until the user has supplied values for both language and name, as shown in the following code and in Figure 8-3:

```
server <- function(input, output, session) {
  greetings <- c(
    English = "Hello",
    Maori = "Ki ora"
  )
  output$greeting <- renderText({
    req(input$language, input$name)
```

```
      paste0(greetings[[input$language]], " ", input$name, "!")
    })
  }
```

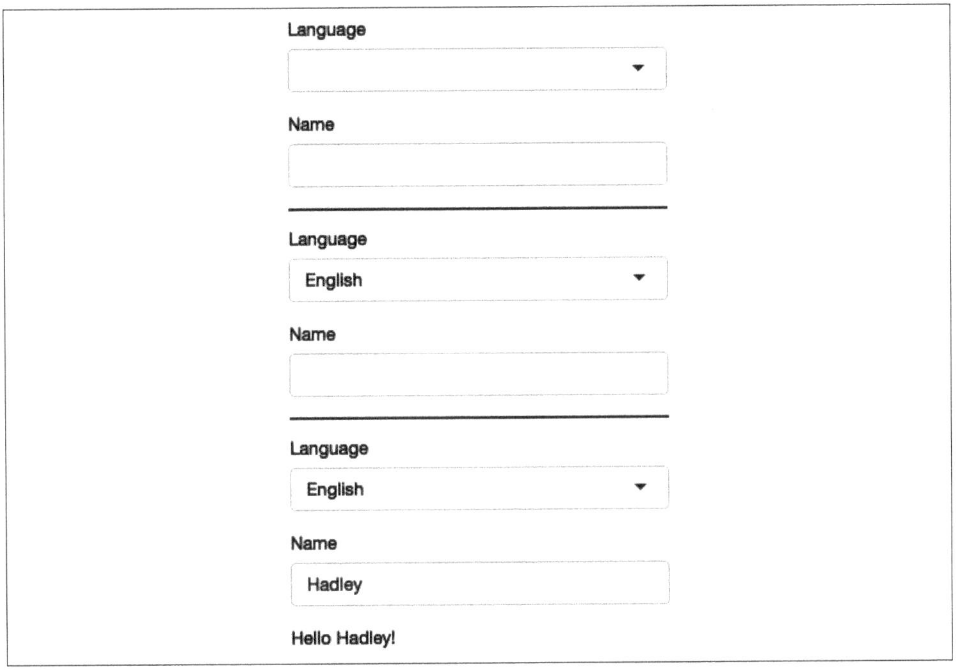

Figure 8-3. *By using* `req()`, *the output is only shown once both language and name have been supplied. See live at https://hadley.shinyapps.io/ms-require-simple2.*

`req()` works by signaling a special *condition*.[1] This special condition causes all downstream reactives and outputs to stop executing. Technically, it leaves any downstream reactive consumers in an invalidated state. We'll come back to this terminology in Chapter 16.

`req()` is designed so that `req(input$x)` will only proceed if the user has supplied a value, regardless of the type of input control.[2] You can also use `req()` with your own logical statement if needed. For example, `req(input$a > 0)` will permit computation to proceed when `a` is greater than 0; this is typically the form you'll use when performing validation, as we'll see next.

---

1 *Condition* is a technical term that includes errors, warnings, and messages. If you're interested, you can learn more of the details of R's condition system in Chapter 8 of *Advanced R* (*https://oreil.ly/gP3i8*).

2 More precisely, `req()` proceeds only if its inputs are *truthy*, that is, any value apart from FALSE, NULL , "", or a handful of other special cases described in `?isTruthy`.

req() and Validation

Let's combine `req()` and shinyFeedback to solve a more challenging problem. I'm going to return to the simple app we made in Chapter 1, which allowed you to select a built-in dataset and see its contents. I'm going to make it more general and more complex by using `textInput()` instead of `selectInput()`. The UI changes very little:

```
ui <- fluidPage(
  shinyFeedback::useShinyFeedback(),
  textInput("dataset", "Dataset name"),
  tableOutput("data")
)
```

But the server function needs to get a little more complex. We're going to use `req()` in two ways:

- We only want to proceed with computation if the user has entered a value, so we do `req(input$dataset)`.
- Then we check to see if the supplied name actually exists. If it doesn't, we display an error message and then use `req()` to cancel computation. Note the use of `can celOutput = TRUE`: normally, canceling a reactive will reset all downstream outputs; using `cancelOutput = TRUE` leaves them displaying the last good value. This is important for `textInput()`, which may trigger an update while you're in the middle of typing a name.

The results are shown in Figure 8-4.

```
server <- function(input, output, session) {
  data <- reactive({
    req(input$dataset)

    exists <- exists(input$dataset, "package:datasets")
    shinyFeedback::feedbackDanger("dataset", !exists, "Unknown dataset")
    req(exists, cancelOutput = TRUE)

    get(input$dataset, "package:datasets")
  })

  output$data <- renderTable({
    head(data())
  })
}
```

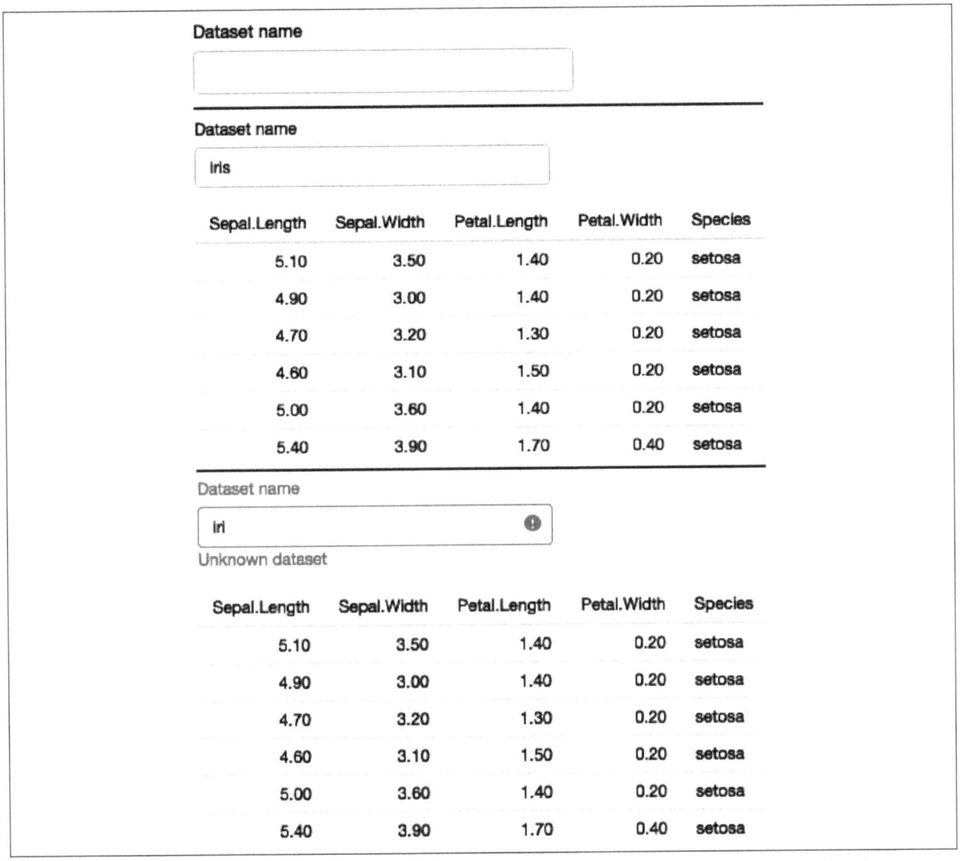

| Dataset name | | | | |
| | | | | |

| Dataset name | | | | |
| iris | | | | |

Sepal.Length	Sepal.Width	Petal.Length	Petal.Width	Species
5.10	3.50	1.40	0.20	setosa
4.90	3.00	1.40	0.20	setosa
4.70	3.20	1.30	0.20	setosa
4.60	3.10	1.50	0.20	setosa
5.00	3.60	1.40	0.20	setosa
5.40	3.90	1.70	0.40	setosa

| Dataset name | | | | |
| iri | | | | ⊕ |

Unknown dataset

Sepal.Length	Sepal.Width	Petal.Length	Petal.Width	Species
5.10	3.50	1.40	0.20	setosa
4.90	3.00	1.40	0.20	setosa
4.70	3.20	1.30	0.20	setosa
4.60	3.10	1.50	0.20	setosa
5.00	3.60	1.40	0.20	setosa
5.40	3.90	1.70	0.40	setosa

Figure 8-4. On load, the table is empty because the dataset name is empty. The data is shown after we type a valid dataset name (iris) and continues to be shown when we press Backspace in order to type a new dataset name. See live at https:// hadley.shinyapps.io/ms-require-cancel.

Validate Output

shinyFeedback is great when the problem is related to a single input. But sometimes the invalid state is a result of a combination of inputs. In this case, it doesn't really make sense to put the error next to an input (which one would you put it beside?), and instead it makes more sense to put it in the output.

You can do so with a tool built into shiny: `validate()`. When called inside a reactive or an output, `validate(message)` stops execution of the rest of the code and instead displays `message` in any downstream outputs. The following code shows a simple example where we don't want to log or square-root negative values:

```
ui <- fluidPage(
  numericInput("x", "x", value = 0),
  selectInput("trans", "transformation",
    choices = c("square", "log", "square-root")
  ),
  textOutput("out")
)

server <- function(input, output, server) {
  output$out <- renderText({
    if (input$x < 0 && input$trans %in% c("log", "square-root")) {
      validate("x can not be negative for this transformation")
    }

    switch(input$trans,
      square = input$x ^ 2,
      "square-root" = sqrt(input$x),
      log = log(input$x)
    )
  })
}
```

You can see the results in Figure 8-5.

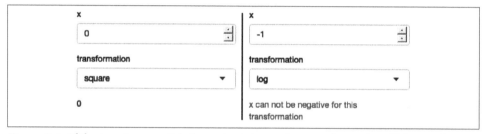

Figure 8-5. If the inputs are valid, the output shows the transformation. If the combination of inputs is invalid, then the output is replaced with an informative message.

Notifications

If there isn't a problem and you just want to let the user know what's happening, use a *notification*. In Shiny, notifications are created with showNotification() and stack in the bottom right of the page. There are three basic ways to use showNotification():

- To show a transient notification that automatically disappears after a fixed amount of time
- To show a notification when a process starts and remove it when the process ends
- To update a single notification with progressive updates

These three techniques are discussed in the following sections.

Transient Notification

The simplest way to use showNotification() is to call it with a single argument: the message that you want to display. It's very hard to capture this behavior with a screenshot, so go to the live app (*https://hadley.shinyapps.io/ms-notification-transient*) if you want to see it in action:

```r
ui <- fluidPage(
  actionButton("goodnight", "Good night")
)
server <- function(input, output, session) {
  observeEvent(input$goodnight, {
    showNotification("So long")
    Sys.sleep(1)
    showNotification("Farewell")
    Sys.sleep(1)
    showNotification("Auf Wiedersehen")
    Sys.sleep(1)
    showNotification("Adieu")
  })
}
```

By default, the message will disappear after five seconds, but you can override it by setting duration, or the user can dismiss it earlier by clicking the close button. If you want to make the notification more prominent, you can set the type argument to one of "message," "warning," or "error":

```r
server <- function(input, output, session) {
  observeEvent(input$goodnight, {
    showNotification("So long")
    Sys.sleep(1)
    showNotification("Farewell", type = "message")
    Sys.sleep(1)
    showNotification("Auf Wiedersehen", type = "warning")
    Sys.sleep(1)
    showNotification("Adieu", type = "error")
  })
}
```

Figure 8-6 gives a sense of what this looks like.

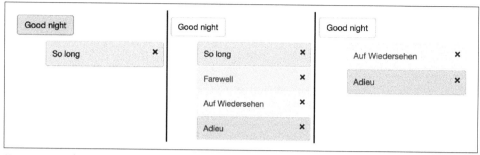

Figure 8-6. The progression of notifications after clicking "Good night": the first notification appears, after three more seconds all notifications are shown, then the notifications start to fade away. See live at https://hadley.shinyapps.io/ms-notify-persistent.

Removing on Completion

It's often useful to tie the presence of a notification to a long-running task. In this case, you want to show the notification when the task starts and remove the notification when the task completes. To do this, you'll need to:

- Set `duration = NULL` and `closeButton = FALSE` so that the notification stays visible until the task is complete.

- Store the `id` returned by `showNotification()`, and then pass this value to `removeNotification()`. The most reliable way to do so is to use `on.exit()`, which ensures that the notification is removed regardless of how the task completes (either successfully or with an error). Read the article "Changing and Restoring State" (*https://oreil.ly/8RFsu*) to learn more about `on.exit()`.

The following example puts the pieces together to show how you might keep the user up to date when reading in a large CSV file:[3]

```
server <- function(input, output, session) {
  data <- reactive({
    id <- showNotification("Reading data...", duration = NULL, closeButton = FALSE)
    on.exit(removeNotification(id), add = TRUE)

    read.csv(input$file$datapath)
  })
}
```

Generally, these sorts of notifications will live in a reactive, because that ensures that the long-running computation is only rerun when needed.

---

[3] If reading CSV files is a bottleneck in your application, you should consider using `data.table::fread()` and `vroom::vroom()` instead; they can be orders of magnitude faster than `read.csv()`.

Progressive Updates

As you saw in the first example, multiple calls to showNotification() usually create multiple notifications. You can instead update a single notification by capturing the id from the first call and using it in subsequent calls:

```r
ui <- fluidPage(
  tableOutput("data")
)

server <- function(input, output, session) {
  notify <- function(msg, id = NULL) {
    showNotification(msg, id = id, duration = NULL, closeButton = FALSE)
  }

  data <- reactive({
    id <- notify("Reading data...")
    on.exit(removeNotification(id), add = TRUE)
    Sys.sleep(1)

    notify("Reticulating splines...", id = id)
    Sys.sleep(1)

    notify("Herding llamas...", id = id)
    Sys.sleep(1)

    notify("Orthogonalizing matrices...", id = id)
    Sys.sleep(1)

    mtcars
  })

  output$data <- renderTable(head(data()))
}
```

This is useful if your long-running task has multiple subcomponents. You can see the results in the live app (*https://hadley.shinyapps.io/ms-notification-updates*).

Progress Bars

For long-running tasks, the best type of feedback is a progress bar. As well as telling you where you are in the process, it also helps you estimate how much longer it's going to be: Should you take a deep breath, go get a coffee, or come back tomorrow? In this section, I'll show two techniques for displaying progress bars, one built into Shiny and one from the waiter (*https://waiter.john-coene.com*) package developed by John Coene.

Unfortunately, both techniques suffer from the same major drawback: to use a progress bar, you need to be able to divide the big task into a known number of small pieces that each take roughly the same amount of time. This is often hard, particularly since the underlying code is often written in C, and it has no way to communicate progress updates to you. We are working on tools in the progress package (*https://github.com/r-lib/progress*) so that packages like dplyr, readr, and vroom will one day generate progress bars that you can easily forward to Shiny.

Shiny

To create a progress bar with Shiny, you need to use `withProgress()` and `incProgress()`. Imagine you have some slow-running code that looks like this:[4]

```
for (i in seq_len(step)) {
  x <- function_that_takes_a_long_time(x)
}
```

You start by wrapping it in `withProgress()`. This shows the progress bar when the code starts and automatically removes it when it's done:

```
withProgress({
  for (i in seq_len(step)) {
    x <- function_that_takes_a_long_time(x)
  }
})
```

Then call `incProgress()` after each step:

```
withProgress({
  for (i in seq_len(step)) {
    x <- function_that_takes_a_long_time(x)
    incProgress(1 / length(step))
  }
})
```

The first argument of `incProgress()` is the amount to increment the progress bar. By default, the progress bar starts at 0 and ends at 1, so the incrementing by 1 divided by the number of steps will ensure that the progress bar is complete at the end of the loop.

---

4 If your code doesn't involve a for loop or an apply/map function, it's going to be very difficult to make a progress bar.

Here's how that might look in a complete Shiny app, as shown in Figure 8-7:

```
ui <- fluidPage(
  numericInput("steps", "How many steps?", 10),
  actionButton("go", "go"),
  textOutput("result")
)

server <- function(input, output, session) {
  data <- eventReactive(input$go, {
    withProgress(message = "Computing random number", {
      for (i in seq_len(input$steps)) {
        Sys.sleep(0.5)
        incProgress(1 / input$steps)
      }
      runif(1)
    })
  })

  output$result <- renderText(round(data(), 2))
}
```

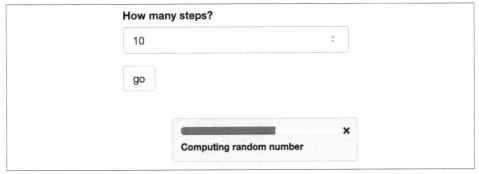

Figure 8-7. A progress bar helps indicate how much longer a calculation has to run. See live at https://hadley.shinyapps.io/ms-progress.

A few things to note:

- I used the optional `message` argument to add some explanatory text to the progress bar.
- I used `Sys.sleep()` to simulate a long-running operation; in your code, this would be a slow function.
- I allowed the user to control when the event starts by combining a button with `eventReactive()`. This is good practice for any task that requires a progress bar.

Waiter

The built-in progress bar is great for the basics, but if you want something that provides more visual options, you might try the waiter (*https://waiter.john-coene.com*) package. Adapting the preceding code to work with waiter is straightforward. In the UI, we add `use_waitress()`:

```r
ui <- fluidPage(
  waiter::use_waitress(),
  numericInput("steps", "How many steps?", 10),
  actionButton("go", "go"),
  textOutput("result")
)
```

The interface for waiter's progress bars are a little different. The waiter package uses an R6 object to bundle all progress-related functions into a single object. If you've never used an R6 object before, don't worry too much about the details; you can just copy and paste this template. The basic life cycle looks like this:

```r
# Create a new progress bar
waitress <- waiter::Waitress$new(max = input$steps)
# Automatically close it when done
on.exit(waitress$close())

for (i in seq_len(input$steps)) {
  Sys.sleep(0.5)
  # increment one step
  waitress$inc(1)
}
```

And we can use it in a Shiny app as follows:

```r
server <- function(input, output, session) {
  data <- eventReactive(input$go, {
    waitress <- waiter::Waitress$new(max = input$steps)
    on.exit(waitress$close())

    for (i in seq_len(input$steps)) {
      Sys.sleep(0.5)
      waitress$inc(1)
    }

    runif(1)
  })

  output$result <- renderText(round(data(), 2))
}
```

The default display is a thin progress bar at the top of the page—which you can see in the live app (*https://hadley.shinyapps.io/ms-waiter*)—but there are a number of ways to customize the output:

- You can override the default `theme` to use one of the following:

  overlay
  > An opaque progress bar that hides the whole page

  overlay-opacity
  > A translucent progress bar that covers the whole page

  overlay-percent
  > An opaque progress bar that also displays a numeric percentage

- Instead of showing a progress bar for the entire page, you can overlay it on an existing input or output by setting the `selector` parameter. For example:

  ```
  waitress <- Waitress$new(selector = "#steps", theme = "overlay")
  ```

Spinners

Sometimes you don't know exactly how long an operation will take, and you just want to display an animated spinner that reassures the user that something is happening. You can also use the waiter package for this task; just switch from using a `Waitress` to using a `Waiter`:

```
ui <- fluidPage(
  waiter::use_waiter(),
  actionButton("go", "go"),
  textOutput("result")
)

server <- function(input, output, session) {
  data <- eventReactive(input$go, {
    waiter <- waiter::Waiter$new()
    waiter$show()
    on.exit(waiter$hide())

    Sys.sleep(sample(5, 1))
    runif(1)
  })
  output$result <- renderText(round(data(), 2))
}
```

Figure 8-8 shows how this will appear in the app.

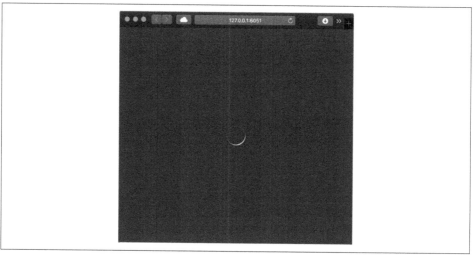

Figure 8-8. A `waiter` displays a whole app spinner while something is happening. See live at https://hadley.shinyapps.io/ms-spinner-1.

Like `Waitress`, you can also use `Waiters` for specific outputs. These `waiters` can automatically remove the spinner when the output updates, so the code is even simpler:

```r
ui <- fluidPage(
  waiter::use_waiter(),
  actionButton("go", "go"),
  plotOutput("plot"),
)

server <- function(input, output, session) {
  data <- eventReactive(input$go, {
    waiter::Waiter$new(id = "plot")$show()

    Sys.sleep(3)
    data.frame(x = runif(50), y = runif(50))
  })

  output$plot <- renderPlot(plot(data()), res = 96)
}
```

Figure 8-9 shows the result.

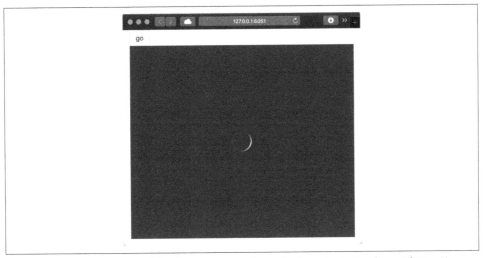

Figure 8-9. You can instead display a spinner for a single output. See live at https://hadley.shinyapps.io/ms-spinner-2.

The waiter package provides a large variety of spinners to choose from; see your options at `?waiter::spinners` and then choose one with (for example) `Waiter $new(html = spin_ripple())`.

An even simpler alternative is to use the shinycssloaders (*https://oreil.ly/qIcuN*) package by Dean Attali. It uses JavaScript to listen to Shiny events, so it doesn't even need any code on the server side. Instead, you just use `shinycssloaders::withSpinner()` to wrap outputs from which you want to automatically get a spinner when they have been invalidated:

```
library(shinycssloaders)

ui <- fluidPage(
  actionButton("go", "go"),
  withSpinner(plotOutput("plot")),
)
server <- function(input, output, session) {
  data <- eventReactive(input$go, {
    Sys.sleep(3)
    data.frame(x = runif(50), y = runif(50))
  })

  output$plot <- renderPlot(plot(data()), res = 96)
}
```

Confirming and Undoing

Sometimes an action is potentially dangerous, and you either want to make sure that the user *really* wants to do it or you want to give them the ability to back out before it's too late. The three techniques in this section lay out your basic options and give you some tips for how you might implement them in your app.

Explicit Confirmation

The simplest approach to protecting the user from accidentally performing a dangerous action is to require an explicit confirmation. The easiest way is to use a dialog box, which forces the user to pick from one of a small set of actions. In Shiny, you create a dialog box with `modalDialog()`. This is called a "modal" dialog because it creates a new "mode" of interaction; you can't interact with the main application until you have dealt with the dialog.

Imagine you have a Shiny app that deletes some files from a directory (or rows in a database, etc.). This is hard to undo, so you want to make sure that the user is really sure. You could create a dialog box, as shown in Figure 8-10, that requires an explicit confirmation, as follows:

```
modal_confirm <- modalDialog(
  "Are you sure you want to continue?",
  title = "Deleting files",
  footer = tagList(
    actionButton("cancel", "Cancel"),
    actionButton("ok", "Delete", class = "btn btn-danger")
  )
)
```

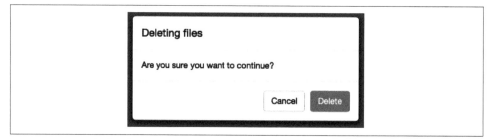

Figure 8-10. A dialog box checks whether or not you want to delete some files.

There are a few small, but important, details to consider when creating a dialog box:

- What should you call the buttons? It's best to be descriptive, so avoid yes/no or continue/cancel in favor of recapitulating the key verb.

- How should you order the buttons? Do you put cancel first (like the Mac) or continue first (like Windows)? Your best option is to mirror the platform that you think most people will be using.

- Can you make the dangerous option more obvious? Here I've used `class = "btn btn-danger"` to style the button prominently.

Jakob Nielsen has more good advice in his "OK-Cancel or Cancel-OK? The Trouble With Buttons" (*https://oreil.ly/php5k*) article.

Let's use this dialog in a real (if very simple) app. Our UI exposes a single button to "delete all the files":

```
ui <- fluidPage(
  actionButton("delete", "Delete all files?")
)
```

There are two new ideas in the `server()`:

- We use `showModal()` and `removeModal()` to show and hide the dialog.

- We observe events generated by the UI from `modal_confirm`. These objects aren't created statically in the `ui` but are instead dynamically added in the `server()` by `showModal()`:

```
server <- function(input, output, session) {
  observeEvent(input$delete, {
    showModal(modal_confirm)
  })

  observeEvent(input$ok, {
    showNotification("Files deleted")
    removeModal()
  })
  observeEvent(input$cancel, {
    removeModal()
  })
}
```

You'll see this idea in much more detail in Chapter 10.

Undoing an Action

Explicit confirmation is most useful for destructive actions that are only performed infrequently. You should avoid it if you want to reduce the errors made by frequent actions. For example, this technique would not work for Twitter. If there was a dialog box that said "Are you sure you want to tweet this?" you would soon learn to automatically click yes and still feel the same feeling of regret when you notice a typo 10 seconds after tweeting.

In this situation, a better approach is to wait a few seconds before actually performing the action, giving the user a chance to notice any problems and undo them. This isn't really an undo (since you're not actually doing anything), but it's an evocative word that users will understand.

I illustrate the technique with a website that I personally wish had an undo button: Twitter. The essence of the Twitter UI is very simple: there's a text area to compose your tweet and a button to send it:

```r
ui <- fluidPage(
  textAreaInput("message",
    label = NULL,
    placeholder = "What's happening?",
    rows = 3
  ),
  actionButton("tweet", "Tweet")
)
```

The server function is quite complex and requires some techniques that we haven't talked about. Don't worry too much about understanding the code; focus on the basic idea: we use some special arguments to observeEvent() to run some code after a few seconds. The big new idea is that we capture the result of observeEvent() and save it to a variable; this allows us to destroy the observer so the code that would really send the tweet is never run:

```r
runLater <- function(action, seconds = 3) {
  observeEvent(
    invalidateLater(seconds * 1000), action,
    ignoreInit = TRUE,
    once = TRUE,
    ignoreNULL = FALSE,
    autoDestroy = FALSE
  )
}

server <- function(input, output, session) {
  waiting <- NULL
  last_message <- NULL

  observeEvent(input$tweet, {
    notification <- glue::glue("Tweeted '{input$message}'")
    last_message <<- input$message
    updateTextAreaInput(session, "message", value = "")

    showNotification(
      notification,
      action = actionButton("undo", "Undo?"),
      duration = NULL,
      closeButton = FALSE,
      id = "tweeted",
      type = "warning"
```

```
    )

    waiting <<- runLater({
      cat("Actually sending tweet...\n")
      removeNotification("tweeted")
    })
  })

  observeEvent(input$undo, {
    waiting$destroy()
    showNotification("Tweet retracted", id = "tweeted")
    updateTextAreaInput(session, "message", value = last_message)
  })
}
```

See it in action in the live app (*https://hadley.shinyapps.io/ms-undo*).

Trash

For actions that you might regret days later, a more sophisticated pattern is to implement something like the trash or recycling bin on your computer. When you delete a file, it isn't permanently deleted but instead is moved to a holding cell, which requires a separate action to empty. This is like the "undo" option on steroids; you have a lot of time to regret your action. It's also a bit like the confirmation; you have to do two separate actions to make deletion permanent.

The primary downside of this technique is that it is substantially more complicated to implement (you have to have a separate "holding cell" that stores the information needed to undo the action) and requires regular intervention from the user to avoid accumulating. For that reason, I think it's beyond the scope of all but the most complicated Shiny apps, so I'm not going to show an implementation here.

Summary

This chapter has given you a number of tools to help communicate to the user what's happening with your app. In some sense, these techniques are mostly optional. But while your app will work without them, their thoughtful application can have a huge impact on the quality of the user experience. You can often omit feedback when you're the only user of an app, but the more people use it, the more that thoughtful notification will pay off.

In the next chapter, you'll learn how to transfer files to and from the user.

## CHAPTER 9

Uploads and Downloads

Transferring files to and from the user is a common feature of apps. You can use it to upload data for analysis or download the results as a dataset or as a report. This chapter shows the UI and server components that you'll need to transfer files in and out of your app. We begin by loading shiny:

```
library(shiny)
```

Upload

We'll start by discussing file uploads, showing you the basic UI and server components, and then showing how they fit together in a simple app.

UI

The UI needed to support file uploads is simple: just add `fileInput()` to your UI:

```
ui <- fluidPage(
  fileInput("upload", "Upload a file")
)
```

Like most other UI components, there are only two required arguments: `id` and `label`. The `width`, `buttonLabel`, and `placeholder` arguments allow you to tweak the appearance in other ways. I won't discuss them here, but you can read more about them in `?fileInput`.

Server

Handling fileInput() on the server is a little more complicated than other inputs. Most inputs return simple vectors, but fileInput() returns a data frame with four columns:

name
> The original filename on the user's computer.

size
> The file size, in bytes. By default, the user can only upload files up to 5 MB. You can increase this limit by setting the shiny.maxRequestSize option prior to starting Shiny—to allow up to 10 MB run options(shiny.maxRequestSize = 10 * 1024^2), for example.

type
> The "MIME type" of the file.[1] This is a formal specification of the file type that is usually derived from the extension and is rarely needed in Shiny apps.

datapath
> The path to where the data has been uploaded on the server. Treat this path as ephemeral: if the user uploads more files, this file may be deleted. The data is always saved to a temporary directory and given a temporary name.

I think the easiest way to understand this data structure is to make a simple app. Run the following code and upload a few files to get a sense of what data Shiny is providing:

```
ui <- fluidPage(
  fileInput("upload", NULL, buttonLabel = "Upload...", multiple = TRUE),
  tableOutput("files")
)
server <- function(input, output, session) {
  output$files <- renderTable(input$upload)
}
```

You can see the results after I uploaded a couple of puppy photos (from "Images" on page 116) in Figure 9-1.

---

1 MIME type is short for "**m**ultipurpose **i**nternet **m**ail **e**xtensions type." As you might guess from the name, it was originally designed for email systems, but now it's used widely across many internet tools. A MIME type looks like type/subtype. Some common examples are text/csv, text/html, image/png, application/pdf, application/vnd.ms-excel (excel file).

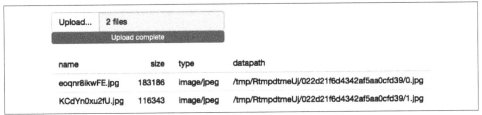

Figure 9-1. This simple app lets you see exactly what data Shiny provides to you for uploaded files. See live at https://hadley.shinyapps.io/ms-upload.

Note my use of the `label` and `buttonLabel` arguments to mildly customize the appearance, and my use of `multiple = TRUE` to allow the user to upload multiple files.

Uploading Data

If the user is uploading a dataset, there are two details that you need to be aware of:

- `input$upload` is initialized to `NULL` on page load, so you'll need `req(input$file)` to make sure your code waits until the first file is uploaded.
- The `accept` argument allows you to limit the possible inputs. The easiest way is to supply a character vector of file extensions, like `accept = ".csv"`. But the `accept` argument is only a suggestion to the browser and is not always enforced, so it's good practice to also validate it (e.g., "Validation" on page 119) yourself. The easiest way to get the file extension in R is `tools::file_ext()`, but just be aware that it removes the leading . from the extension.

Putting all these ideas together gives us the following, where you can upload a .csv or .tsv file and see the first n rows:

```r
ui <- fluidPage(
  fileInput("file", NULL, accept = c(".csv", ".tsv")),
  numericInput("n", "Rows", value = 5, min = 1, step = 1),
  tableOutput("head")
)

server <- function(input, output, session) {
  data <- reactive({
    req(input$file)

    ext <- tools::file_ext(input$file$name)
    switch(ext,
      csv = vroom::vroom(input$file$datapath, delim = ","),
      tsv = vroom::vroom(input$file$datapath, delim = "\t"),
      validate("Invalid file; Please upload a .csv or .tsv file")
    )
```

```
  })
  output$head <- renderTable({
    head(data(), input$n)
  })
}
```

See it in action in the live app (*https://hadley.shinyapps.io/ms-upload-validate*).

Note that since multiple = FALSE (the default), input$file will be a single-row data frame, and input$file$name and input$file$datapath will be a length-1 character vector.

Download

Next, we'll look at file downloads, showing you the basic UI and server components, then demonstrating how you might use them to allow the user to download data or reports.

Basics

Again, the UI is straightforward: use downloadButton(id) or downloadLink(id) to give the user something to click to download a file:

```
ui <- fluidPage(
  downloadButton("download1"),
  downloadLink("download2")
)
```

The results are shown in Figure 9-2.

Figure 9-2. A download button and a download link.

You can customize their appearance using the same class and icon arguments as for actionButtons(), as described in "Action Buttons" on page 20.

Unlike other outputs, downloadButton() is not paired with a render function. Instead, you use downloadHandler(), which looks something like this:

```
output$download <- downloadHandler(
  filename = function() {
    paste0(input$dataset, ".csv")
  },
  content = function(file) {
    write.csv(data(), file)
  }
)
```

`downloadHandler()` has two arguments, both functions:

`filename`
> A function with no arguments that returns a filename (as a string). The job of this function is to create the name that will be shown to the user in the download dialog box.

`content`
> A function with one argument, `file`, which is the path to save the file. The job of this function is to save the file in a place that Shiny knows about so it can then send it to the user.

This is an unusual interface, but it allows Shiny to control where the file should be saved (so it can be placed in a secure location) while you still control the contents of that file.

Next we'll put these pieces together to show how to transfer data files or reports to the user.

Downloading Data

The following app shows off the basics of data download by allowing you to download any dataset in the datasets package as a tab-separated file, as shown in Figure 9-3. I recommend using `.tsv` (tab-separated values) instead of `.csv` (comma-separated values) because many European countries use commas to separate the whole and fractional parts of a number (e.g., `1,23` versus `1.23`). This means they can't use commas to separate fields and instead use semicolons in so-called CSV files! You can avoid this complexity by using tab-separated files, which work the same way everywhere:

```r
ui <- fluidPage(
  selectInput("dataset", "Pick a dataset", ls("package:datasets")),
  tableOutput("preview"),
  downloadButton("download", "Download .tsv")
)

server <- function(input, output, session) {
  data <- reactive({
    out <- get(input$dataset, "package:datasets")
    if (!is.data.frame(out)) {
      validate(paste0("'", input$dataset, "' is not a data frame"))
    }
    out
  })

  output$preview <- renderTable({
    head(data())
  })
```

```
output$download <- downloadHandler(
  filename = function() {
    paste0(input$dataset, ".tsv")
  },
  content = function(file) {
    vroom::vroom_write(data(), file)
  }
)
}
```

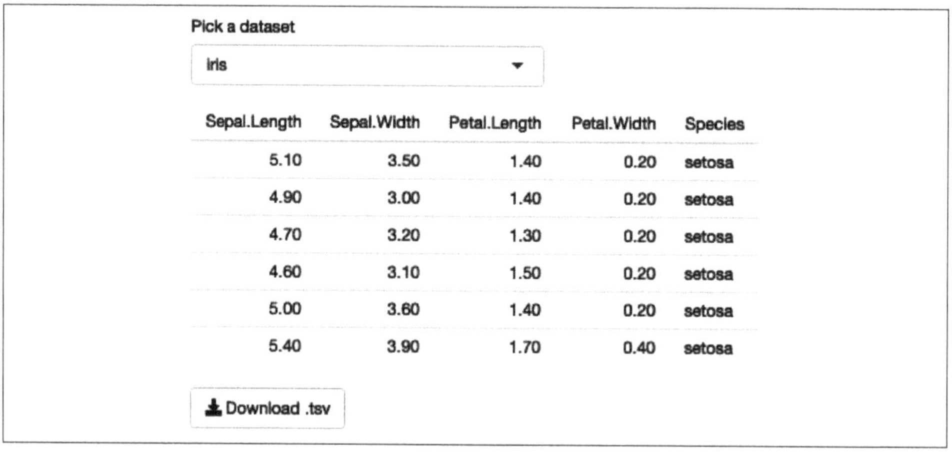

Pick a dataset

iris ▼

Sepal.Length	Sepal.Width	Petal.Length	Petal.Width	Species
5.10	3.50	1.40	0.20	setosa
4.90	3.00	1.40	0.20	setosa
4.70	3.20	1.30	0.20	setosa
4.60	3.10	1.50	0.20	setosa
5.00	3.60	1.40	0.20	setosa
5.40	3.90	1.70	0.40	setosa

⬇ Download .tsv

Figure 9-3. This richer app allows you to select a built-in dataset and preview it before downloading. See live at https://hadley.shinyapps.io/ms-download-data.

Note the use of `validate()` to only allow the user to download datasets that are data frames. A better approach would be to prefilter the list, but this lets you see another application of `validate()`.

Downloading Reports

As well as downloading data, you may want the users of your app to download a report that summarizes the result of interactive exploration in the Shiny app. This is quite a lot of work, because you also need to display the same information in a different format, but it is very useful for high-stakes apps.

One powerful way to generate such a report is with a parameterized RMarkdown document (*https://oreil.ly/ZlvJg*). A parameterized RMarkdown file has a `params` field in the YAML metadata:

```
title: My Document
output: html_document
params:
  year: 2018
  region: Europe
```

```
printcode: TRUE
data: file.csv
```

Inside the document, you can refer to these values using elements of the `params` list (e.g., `params$year`, `params$region`). The values in the YAML metadata are defaults; you'll generally override them by providing the `params` argument in a call to `rmarkdown::render()`. This makes it easy to generate many different reports from the same *.Rmd*.

Here's a simple example adapted from "Generating downloadable reports" (*https:// oreil.ly/QcleZ*), which describes this technique in more detail. The key idea is to call `rmarkdown::render()` from the `content` argument of `downloadHandler()`:

```
ui <- fluidPage(
  sliderInput("n", "Number of points", 1, 100, 50),
  downloadButton("report", "Generate report")
)

server <- function(input, output, session) {
  output$report <- downloadHandler(
    filename = "report.html",
    content = function(file) {
      params <- list(n = input$n)

      id <- showNotification(
        "Rendering report...",
        duration = NULL,
        closeButton = FALSE
      )
      on.exit(removeNotification(id), add = TRUE)

      rmarkdown::render("report.Rmd",
        output_file = file,
        params = params,
        envir = new.env(parent = globalenv())
      )
    }
  )
}
```

If you want to produce other output formats, just change the output format in the *.Rmd*, and make sure to update the extension (e.g., to .pdf). See it in action in the live app (*https://hadley.shinyapps.io/ms-download-rmd*). It'll generally take at least a few seconds to render an *.Rmd*, so this is a good place to use a notification from "Notifications" on page 126.

There are a couple of other tricks worth knowing about:

- RMarkdown works in the current working directory, which will fail in many deployment scenarios (e.g., on shinyapps.io). You can work around this by copying the report to a temporary directory when your app starts (i.e., outside of the server function):

```
report_path <- tempfile(fileext = ".Rmd")
file.copy("report.Rmd", report_path, overwrite = TRUE)
```

Then replace `"report.Rmd"` with `report_path` in the call to `rmarkdown::render()`:

```
rmarkdown::render(report_path,
  output_file = file,
  params = params,
  envir = new.env(parent = globalenv())
)
```

- By default, RMarkdown will render the report in the current process, which means that it will inherit many settings from the Shiny app (like loaded packages, options, etc.). For greater robustness, I recommend running `render()` in a separate R session using the callr package:

```
render_report <- function(input, output, params) {
  rmarkdown::render(input,
    output_file = output,
    params = params,
    envir = new.env(parent = globalenv())
  )
}

server <- function(input, output) {
  output$report <- downloadHandler(
    filename = "report.html",
    content = function(file) {
      params <- list(n = input$slider)
      callr::r(
        render_report,
        list(input = report_path, output = file, params = params)
      )
    }
  )
}
```

You can see all these pieces put together in `rmarkdown-report/` (*https://oreil.ly/OZuFs*), found inside the Mastering Shiny GitHub repo.

The shinymeta (*https://oreil.ly/f8uj4*) package solves a related problem: sometimes you need to be able to turn the current state of a Shiny app into a reproducible report that can be rerun in the future. Joe Cheng's useR! 2019 keynote "Shiny's Holy Grail: Interactivity with Reproducibility" (*https://oreil.ly/Q7f2P*) offers more information.

Case Study

To finish up, we'll work through a small case study where we upload a file (with user-supplied separator), preview it, perform some optional transformations using the janitor package (*http://sfirke.github.io/janitor*), by Sam Firke, and then let the user download it as a .tsv.

To make it easier to understand how to use the app, I've used sidebarLayout() to divide the app into three main steps:

1. Uploading and parsing the file:

   ```
   ui_upload <- sidebarLayout(
     sidebarPanel(
       fileInput("file", "Data", buttonLabel = "Upload..."),
       textInput("delim", "Delimiter (leave blank to guess)", ""),
       numericInput("skip", "Rows to skip", 0, min = 0),
       numericInput("rows", "Rows to preview", 10, min = 1)
     ),
     mainPanel(
       h3("Raw data"),
       tableOutput("preview1")
     )
   )
   ```

2. Cleaning the file:

   ```
   ui_clean <- sidebarLayout(
     sidebarPanel(
       checkboxInput("snake", "Rename columns to snake case?"),
       checkboxInput("constant", "Remove constant columns?"),
       checkboxInput("empty", "Remove empty cols?")
     ),
     mainPanel(
       h3("Cleaner data"),
       tableOutput("preview2")
     )
   )
   ```

3. Downloading the file:

   ```
   ui_download <- fluidRow(
     column(width = 12, downloadButton("download", class = "btn-block"))
   )
   ```

These get assembled into a single `fluidPage()`:

```
ui <- fluidPage(
  ui_upload,
  ui_clean,
  ui_download
)
```

This same organization makes it easier to understand the app:

```
server <- function(input, output, session) {
  # Upload ----------------------------------------------------------
  raw <- reactive({
    req(input$file)
    delim <- if (input$delim == "") NULL else input$delim
    vroom::vroom(input$file$datapath, delim = delim, skip = input$skip)
  })
  output$preview1 <- renderTable(head(raw(), input$rows))

  # Clean -----------------------------------------------------------
  tidied <- reactive({
    out <- raw()
    if (input$snake) {
      names(out) <- janitor::make_clean_names(names(out))
    }
    if (input$empty) {
      out <- janitor::remove_empty(out, "cols")
    }
    if (input$constant) {
      out <- janitor::remove_constant(out)
    }

    out
  })
  output$preview2 <- renderTable(head(tidied(), input$rows))

  # Download --------------------------------------------------------
  output$download <- downloadHandler(
    filename = function() {
      paste0(tools::file_path_sans_ext(input$file$name), ".tsv")
    },
    content = function(file) {
      vroom::vroom_write(tidied(), file)
    }
  )
}
```

Figure 9-4 displays the result.

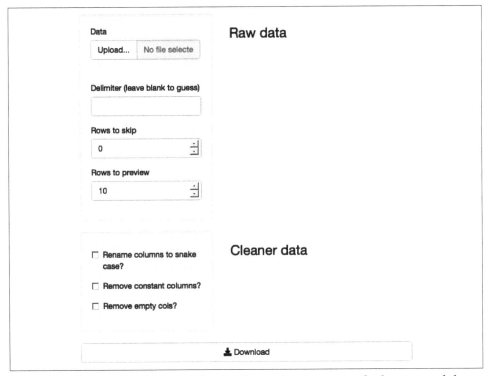

Figure 9-4. This app lets the user upload a file, perform some simple cleaning, and then download the results. See live at https://hadley.shinyapps.io/ms-case-study.

Exercises

1. Use the ambient (*https://ambient.data-imaginist.com*) package by Thomas Lin Pedersen to generate worley noise (*https://oreil.ly/qA10s*) and download a PNG of it.

2. Create an app that lets you upload a CSV file, select a variable, and then perform a `t.test()` on that variable. After the user has uploaded the CSV file, you'll need to use `updateSelectInput()` to fill in the available variables. See "Updating Inputs" on page 153 for details.

3. Create an app that lets the user upload a CSV file, select one variable, draw a histogram, and then download the histogram. For an additional challenge, allow the user to select from `.png`, `.pdf`, and `.svg` output formats.

4. Write an app that allows the user to create a Lego mosaic from any `.png` file using Ryan Timpe's brickr (*https://github.com/ryantimpe/brickr*) package. Once you've completed the basics, add controls to allow the user to select the size of the

mosaic (in bricks), and choose whether to use "universal" or "generic" color palettes.

5. The final app in "Case Study" on page 149 contains this one large reactive:

```
tidied <- reactive({
  out <- raw()
  if (input$snake) {
    names(out) <- janitor::make_clean_names(names(out))
  }
  if (input$empty) {
    out <- janitor::remove_empty(out, "cols")
  }
  if (input$constant) {
    out <- janitor::remove_constant(out)
  }

  out
})
```

Break it up into multiple pieces so that (for example) `jani tor::make_clean_names()` is not rerun when `input$empty` changes.

Summary

In this chapter, you've learned how to transfer files to and from the user using `fileIn put()` and `downloadButton()`. Most of the challenges arise either handling the uploaded files or generating the files to download, so I showed you how to handle a couple of common cases. If I didn't cover your specific challenge here, you'll need to apply your own unique creativity to the problem 😄.

The next chapter will help you handle a common challenge when working with user supplied data: you need to dynamically adapt the user interface to better fit the data. I'll start with some simple techniques that are easy to understand and can be applied in many situations, and gradually work my way up to fully a dynamic user interface generated by code.

## CHAPTER 10

Dynamic UI

So far, we've seen a clean separation between the user interface and the server function: the user interface is defined statically when the app is launched so it can't respond to anything that happens in the app. In this chapter, you'll learn how to create *dynamic* user interfaces, changing the UI using code run in the server function.

There are three key techniques for creating dynamic user interfaces:

- Using the `update` family of functions to modify parameters of input controls
- Using `tabsetPanel()` to conditionally show and hide parts of the user interface
- Using `uiOutput()` and `renderUI()` to generate selected parts of the user interface with code

These three tools give you considerable power to respond to the user by modifying inputs and outputs. I'll demonstrate some of the more useful ways in which you can apply them, but ultimately you're only constrained by your creativity. At the same time, these tools can make your app substantially more difficult to understand, so deploy them sparingly, and always strive to use the simplest technique that solves your problem. Let's begin:

```
library(shiny)
library(dplyr, warn.conflicts = FALSE)
```

Updating Inputs

We'll start with a simple technique that allows you to modify an input after it has been created: the update family of functions. Every input control—for example `tex tInput()`—is paired with an *update function*—for example `updateTextInput()`—that allows you to modify the control after it has been created.

Take the example in the following code, with the results shown in Figure 10-1. The app has two inputs that control the range (the min and max) of another input, a slider. The key idea is to use observeEvent()[1] to trigger updateSliderInput() whenever the min or max inputs change:

```
ui <- fluidPage(
  numericInput("min", "Minimum", 0),
  numericInput("max", "Maximum", 3),
  sliderInput("n", "n", min = 0, max = 3, value = 1)
)
server <- function(input, output, session) {
  observeEvent(input$min, {
    updateSliderInput(inputId = "n", min = input$min)
  })
  observeEvent(input$max, {
    updateSliderInput(inputId = "n", max = input$max)
  })
}
```

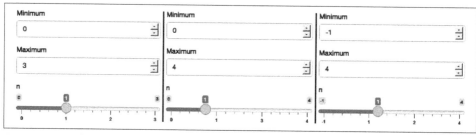

Figure 10-1. The app on load (left), after increasing max (middle), and then decreasing min (right). See live at https://hadley.shinyapps.io/ms-update-basics.

The update functions look a little different than other Shiny functions: they all take the name of the input (as a string) as the inputId argument.[2] The remaining arguments correspond to the arguments to the input constructor that can be modified after creation.

To help you get the hang of the update functions, I'll show a couple more simple examples, then we'll dive into a more complicated case study using hierarchical select boxes, and finish off by discussing the problem of circular references.

---

1 I introduced observeEvent() in "Observers" on page 49 and will discuss it in more detail in "Observers and Outputs" on page 226.

2 The first argument, session, exists for backward compatibility but is very rarely needed.

Simple Uses

The simplest uses of the update functions are to provide small conveniences for the user. For example, maybe you want to make it easy to reset parameters back to their initial value. The following snippet shows how you might combine an `actionButton()`, `observeEvent()`, and `updateSliderInput()`, with the output shown in Figure 10-2:

```r
ui <- fluidPage(
  sliderInput("x1", "x1", 0, min = -10, max = 10),
  sliderInput("x2", "x2", 0, min = -10, max = 10),
  sliderInput("x3", "x3", 0, min = -10, max = 10),
  actionButton("reset", "Reset")
)

server <- function(input, output, session) {
  observeEvent(input$reset, {
    updateSliderInput(inputId = "x1", value = 0)
    updateSliderInput(inputId = "x2", value = 0)
    updateSliderInput(inputId = "x3", value = 0)
  })
}
```

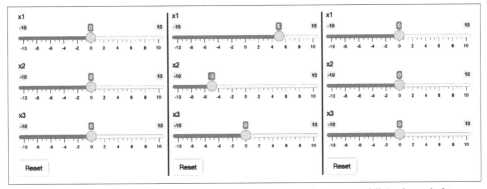

Figure 10-2. The app on load (left), after dragging some sliders (middle), then clicking reset (right). See live at https://hadley.shinyapps.io/ms-update-reset.

A similar application is to tweak the text of an action button so you know exactly what it's going to do. Figure 10-3 shows the results of the following code:

```r
ui <- fluidPage(
  numericInput("n", "Simulations", 10),
  actionButton("simulate", "Simulate")
)

server <- function(input, output, session) {
  observeEvent(input$n, {
    label <- paste0("Simulate ", input$n, " times")
    updateActionButton(inputId = "simulate", label = label)
```

```
    })
  }
```

Figure 10-3. The app on load (left), after setting simulations to 1 (middle), then setting simulations to 100 (right). See live at https://hadley.shinyapps.io/ms-update-button.

There are many ways to use update functions in this way; be on the lookout for ways to give more information to the user when you are working on sophisticated applications. A particularly important application is making it easier to select from a long list of possible options through step-by-step filtering. That's often a problem for "hierarchical select boxes."

Hierarchical Select Boxes

A more complicated, but particularly useful, application of the update functions is to allow interactive drill-down across multiple categories. I'll illustrate their usage with some imaginary data for a sales dashboard that comes from Kaggle (*https://oreil.ly/Oclev*):

```
sales <- vroom::vroom(
  "sales-dashboard/sales_data_sample.csv",
  col_types = list(),
  na = ""
)
sales %>%
  select(TERRITORY, CUSTOMERNAME, ORDERNUMBER, everything()) %>%
  arrange(ORDERNUMBER)
#> # A tibble: 2,823 x 25
#>    TERRITORY CUSTOMERNAME  ORDERNUMBER QUANTITYORDERED PRICEEACH ORDERLINENUMBER
#>    <chr>     <chr>               <dbl>           <dbl>     <dbl>           <dbl>
#>  1 NA        Online Diecas…      10100              30       100               3
#>  2 NA        Online Diecas…      10100              50      67.8               2
#>  3 NA        Online Diecas…      10100              22      86.5               4
#>  4 NA        Online Diecas…      10100              49      34.5               1
#>  5 EMEA      Blauer See Au…      10101              25       100               4
#>  6 EMEA      Blauer See Au…      10101              26       100               1
#>  7 EMEA      Blauer See Au…      10101              45      31.2               3
#>  8 EMEA      Blauer See Au…      10101              46      53.8               2
#>  9 NA        Vitachrome In…      10102              39       100               2
#> 10 NA        Vitachrome In…      10102              41      50.1               1
#> # … with 2,813 more rows, and 19 more variables: SALES <dbl>, ORDERDATE <chr>,
#> #   STATUS <chr>, QTR_ID <dbl>, MONTH_ID <dbl>, YEAR_ID <dbl>,
#> #   PRODUCTLINE <chr>, MSRP <dbl>, PRODUCTCODE <chr>, PHONE <chr>,
#> #   ADDRESSLINE1 <chr>, ADDRESSLINE2 <chr>, CITY <chr>, STATE <chr>,
#> #   POSTALCODE <chr>, COUNTRY <chr>, CONTACTLASTNAME <chr>,
#> #   CONTACTFIRSTNAME <chr>, DEALSIZE <chr>
```

For this demo, I'm going to focus on a natural hierarchy in the data:

- Each territory contains customers.
- Each customer has multiple orders.
- Each order contains rows.

I want to create a user interface where you can:

- Select a territory to see all customers.
- Select a customer to see all orders.
- Select an order to see the underlying rows.

The essence of the UI is simple: I'll create three select boxes and one output table. The choices for the `customername` and `ordernumber` select boxes will be dynamically generated, so I set `choices = NULL`:

```
ui <- fluidPage(
  selectInput("territory", "Territory", choices = unique(sales$TERRITORY)),
  selectInput("customername", "Customer", choices = NULL),
  selectInput("ordernumber", "Order number", choices = NULL),
  tableOutput("data")
)
```

In the server function, I work top-down:

1. I create a reactive, `territory()`, that contains the rows from `sales` that match the selected territory.
2. Whenever `territory()` changes, I update the list of `choices` in the `input$customername` select box.
3. I create another reactive, `customer()`, that contains the rows from `territory()` that match the selected customer.
4. Whenever `customer()` changes, I update the list of `choices` in the `input$ordernumber` select box.
5. I display the selected orders in `output$data`.

You can see that organization here and see the result in Figure 10-4:

```
server <- function(input, output, session) {
  territory <- reactive({
    filter(sales, TERRITORY == input$territory)
  })
  observeEvent(territory(), {
    choices <- unique(territory()$CUSTOMERNAME)
    updateSelectInput(inputId = "customername", choices = choices)
  })
```

```
customer <- reactive({
  req(input$customername)
  filter(territory(), CUSTOMERNAME == input$customername)
})
observeEvent(customer(), {
  choices <- unique(customer()$ORDERNUMBER)
  updateSelectInput(inputId = "ordernumber", choices = choices)
})

output$data <- renderTable({
  req(input$ordernumber)
  customer() %>%
    filter(ORDERNUMBER == input$ordernumber) %>%
    select(QUANTITYORDERED, PRICEEACH, PRODUCTCODE)
})
}
```

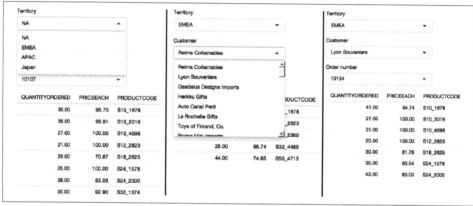

Figure 10-4. I select "EMEA" (left), then "Lyon Souveniers" (middle), then (right) look at the orders. See live at https://hadley.shinyapps.io/ms-update-nested.

Try out the simple app or see the source (*https://oreil.ly/XBU1r*) for a more fully fleshed out application.

Freezing Reactive Inputs

Sometimes this sort of hierarchical selection can briefly create an invalid set of inputs, leading to a flicker of undesirable output. For example, consider this simple app where you select a dataset and then select a variable to summarize:

```
ui <- fluidPage(
  selectInput("dataset", "Choose a dataset", c("pressure", "cars")),
  selectInput("column", "Choose column", character(0)),
  verbatimTextOutput("summary")
)
```

```
server <- function(input, output, session) {
  dataset <- reactive(get(input$dataset, "package:datasets"))

  observeEvent(input$dataset, {
    updateSelectInput(inputId = "column", choices = names(dataset()))
  })

  output$summary <- renderPrint({
    summary(dataset()[[input$column]])
  })
}
```

If you try out the live app (*https://hadley.shinyapps.io/ms-freeze*), you'll notice that when you switch datasets, the summary output will briefly flicker. The reason for this is that updateSelectInput() only has an effect after all outputs and observers have run, so there's temporarily a state where you have dataset B and a variable from dataset A so that the output contains summary(NULL).

You can resolve this problem by "freezing" the input with freezeReactiveValue(). This ensures that any reactives or outputs that use the input won't be updated until the next full round of invalidation:[3]

```
server <- function(input, output, session) {
  dataset <- reactive(get(input$dataset, "package:datasets"))

  observeEvent(input$dataset, {
    freezeReactiveValue(input, "column")
    updateSelectInput(inputId = "column", choices = names(dataset()))
  })

  output$summary <- renderPrint({
    summary(dataset()[[input$column]])
  })
}
```

Note that there's no need to "thaw" the input value; this happens automatically after Shiny detects that the session and server are once again in sync.

You might wonder when you should use freezeReactiveValue(): it's actually good practice to *always* use it when you dynamically change an input value. The actual modification takes some time to flow to the browser then back to Shiny, and in the interim, any reads of the value are at best wasted and at worst lead to errors. Use freezeReactiveValue() to tell all downstream calculations that an input value is stale and that they should save their effort until it's useful.

---

3 To be more precise, any attempt to read a frozen input will result in req(FALSE).

Circular References

There's an important issue we need to discuss if you want to use the update functions to change the current value[4] of an input. From Shiny's perspective, using an update function to modify value is no different from the user modifying the value by clicking or typing. That means an update function can trigger reactive updates in exactly the same way that a human can. This means that you are now stepping outside of the bounds of pure reactive programming, and you need to start worrying about circular references and infinite loops.

For example, take the following simple app. It contains a single input control and an observer that increments its value by one. Every time updateNumericInput() runs, it changes input$n, causing updateNumericInput() to run again, so the app gets stuck in an infinite loop constantly increasing the value of input$n:

```
ui <- fluidPage(
  numericInput("n", "n", 0)
)
server <- function(input, output, session) {
  observeEvent(input$n,
    updateNumericInput(inputId = "n", value = input$n + 1)
  )
}
```

You're unlikely to create such an obvious problem in your own app, but it can crop up if you update multiple controls that depend on one another, as in the next example.

Interrelated Inputs

One place where it's easy to end up with circular references is when you have multiple "sources of truth" in an app. For example, imagine that you want to create a temperature conversion app where you can either enter the temperature in Celsius or in Fahrenheit:

```
ui <- fluidPage(
  numericInput("temp_c", "Celsius", NA, step = 1),
  numericInput("temp_f", "Fahrenheit", NA, step = 1)
)

server <- function(input, output, session) {
  observeEvent(input$temp_f, {
    c <- round((input$temp_f - 32) * 5 / 9)
    updateNumericInput(inputId = "temp_c", value = c)
  })
```

---

4 This is generally only a concern when you are changing the value, but beware that some other parameters can change the value indirectly. For example, if you modify the choices in selectInput() or min and max in sliderInput(), the current value will be modified if it's no longer in the allowed set of values.

```
  observeEvent(input$temp_c, {
    f <- round((input$temp_c * 9 / 5) + 32)
    updateNumericInput(inputId = "temp_f", value = f)
  })
}
```

If you play around with this app (*https://hadley.shinyapps.io/ms-temperature*), you'll notice that it *mostly* works, but you might notice that it'll sometimes trigger multiple changes. For example:

- Set 120 F, then click the down arrow.
- F changes to 119, and C is updated to 48.
- 48 C converts to 118 F, so F changes again to 118.
- Fortunately 118 F is still 48 C, so the updates stop there.

There's no way around this problem because you have one idea (the temperature) with two expressions in the app (Celsius and Fahrenheit). Here we are lucky that cycle quickly converges to a value that satisfies both constraints. In general, you are better off avoiding these situations, unless you are willing to very carefully analyze the convergence properties of the underlying dynamic system that you've created.

Exercises

1. Complete the following user interface with a server function that updates input $date so that you can only select dates in input$year:

   ```
   ui <- fluidPage(
     numericInput("year", "year", value = 2020),
     dateInput("date", "date")
   )
   ```

2. Complete the following user interface with a server function that updates input $county choices based on input$state. For an added challenge, also change the label from "County" to "Parish" for Louisiana and "Borough" for Alaska:

   ```
   library(openintro, warn.conflicts = FALSE)
   states <- unique(county$state)

   ui <- fluidPage(
     selectInput("state", "State", choices = states),
     selectInput("county", "County", choices = NULL)
   )
   ```

3. Complete the following user interface with a server function that updates input $country choices based on the input$continent. Use output$data to display all matching rows:

```
library(gapminder)
continents <- unique(gapminder$continent)

ui <- fluidPage(
  selectInput("continent", "Continent", choices = continents),
  selectInput("country", "Country", choices = NULL),
  tableOutput("data")
)
```

4. Extend the previous app so that you can also choose to select all continents and hence see all countries. You'll need to add "(All)" to the list of choices and then handle that specially when filtering.

5. What is at the heart of the problem described in this RStudio Community post (*https://oreil.ly/WIuFK*)?

Dynamic Visibility

The next step up in complexity is to selectively show and hide parts of the UI. There are more sophisticated approaches if you know a little JavaScript and CSS, but there's a useful technique that doesn't require any extra knowledge: concealing optional UI with a tabset (as introduced in "Tabsets" on page 94). This is a clever hack that allows you to show and hide UI as needed, without having to regenerate it from scratch (as you'll learn in the next section):

```
ui <- fluidPage(
  sidebarLayout(
    sidebarPanel(
      selectInput("controller", "Show", choices = paste0("panel", 1:3))
    ),
    mainPanel(
      tabsetPanel(
        id = "switcher",
        type = "hidden",
        tabPanelBody("panel1", "Panel 1 content"),
        tabPanelBody("panel2", "Panel 2 content"),
        tabPanelBody("panel3", "Panel 3 content")
      )
    )
  )
)

server <- function(input, output, session) {
  observeEvent(input$controller, {
    updateTabsetPanel(inputId = "switcher", selected = input$controller)
  })
}
```

Figure 10-5 shows the result.

Figure 10-5. Selecting panel1 (left), then panel2 (middle), then panel3 (right). See live at https://hadley.shinyapps.io/ms-dynamic-panels.

There are two main ideas here:

- Use tabset panel with hidden tabs.
- Use `updateTabsetPanel()` to switch tabs from the server.

This is a simple idea, but when combined with a little creativity, it gives you a considerable amount of power. The following two sections illustrate a couple of small examples of how you might use it in practice.

Conditional UI

Imagine that you want an app that allows the user to simulate from the normal, uniform, and exponential distributions. Each distribution has different parameters, so we'll need some way to show different controls for different distributions. Here, I'll put the unique user interface for each distribution in its own `tabPanel()` and then arrange the three tabs into a `tabsetPanel()`:

```
parameter_tabs <- tabsetPanel(
  id = "params",
  type = "hidden",
  tabPanel("normal",
    numericInput("mean", "mean", value = 1),
    numericInput("sd", "standard deviation", min = 0, value = 1)
  ),
  tabPanel("uniform",
    numericInput("min", "min", value = 0),
    numericInput("max", "max", value = 1)
  ),
  tabPanel("exponential",
    numericInput("rate", "rate", value = 1, min = 0),
  )
)
```

I'll then embed that inside a fuller UI, which allows the user to pick the number of samples and shows a histogram of the results:

```
ui <- fluidPage(
  sidebarLayout(
    sidebarPanel(
      selectInput("dist", "Distribution",
        choices = c("normal", "uniform", "exponential")
      ),
      numericInput("n", "Number of samples", value = 100),
      parameter_tabs,
    ),
    mainPanel(
      plotOutput("hist")
    )
  )
)
```

Note that I've carefully matched the choices in input$dist to the names of the tab panels. That makes it easy to write the following observeEvent() code that automatically switches controls when the distribution changes. The rest of the app uses techniques that you're already familiar with:

```
server <- function(input, output, session) {
  observeEvent(input$dist, {
    updateTabsetPanel(inputId = "params", selected = input$dist)
  })

  sample <- reactive({
    switch(input$dist,
      normal = rnorm(input$n, input$mean, input$sd),
      uniform = runif(input$n, input$min, input$max),
      exponential = rexp(input$n, input$rate)
    )
  })
  output$hist <- renderPlot(hist(sample()), res = 96)
}
```

See the final result in Figure 10-6. Note that note the value of (for example) input $mean is independent of whether or not it's visible to the user. The underlying HTML control still exists; you just can't see it.

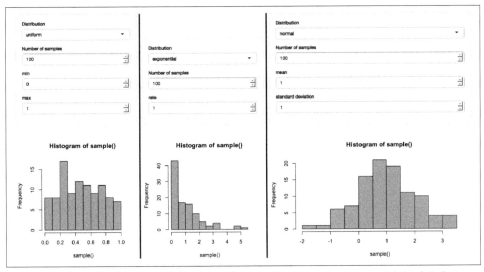

Figure 10-6. Results for normal (left), uniform (middle), and exponential (right) distributions. See live at https://hadley.shinyapps.io/ms-dynamic-conditional.

Wizard Interface

You can also use this idea to create a "wizard," a type of interface that makes it easier to collect a bunch of information by spreading it across multiple pages. Here we embed action buttons within each "page," making it easy to go forward and back:

```r
ui <- fluidPage(
  tabsetPanel(
    id = "wizard",
    type = "hidden",
    tabPanel("page_1",
      "Welcome!",
      actionButton("page_12", "next")
    ),
    tabPanel("page_2",
      "Only one page to go",
      actionButton("page_21", "prev"),
      actionButton("page_23", "next")
    ),
    tabPanel("page_3",
      "You're done!",
      actionButton("page_32", "prev")
    )
  )
)

server <- function(input, output, session) {
  switch_page <- function(i) {
    updateTabsetPanel(inputId = "wizard", selected = paste0("page_", i))
```

```
    }

    observeEvent(input$page_12, switch_page(2))
    observeEvent(input$page_21, switch_page(1))
    observeEvent(input$page_23, switch_page(3))
    observeEvent(input$page_32, switch_page(2))
  }
```

The results are shown in Figure 10-7.

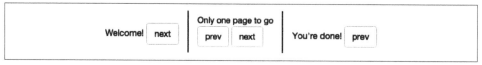

Figure 10-7. A wizard interface portions complex UI over multiple pages. Here we demonstrate the idea with a very simple example, clicking "next" to advance to the next page. See live at https://hadley.shinyapps.io/ms-wizard.

Note the use of the `switch_page()` function to reduce the amount of duplication in the server code. We'll come back to this idea in Chapter 18 and then create a module to automate wizard interfaces in "Wizard" on page 278.

Exercises

1. Use a hidden tabset to show additional controls only if the user checks an "advanced" checkbox.

2. Create an app that plots `ggplot(diamonds, aes(carat))` but allows the user to choose which geom to use: `geom_histogram()`, `geom_freqpoly()`, or `geom_density()`. Use a hidden tabset to allow the user to select different arguments depending on the geom: `geom_histogram()` and `geom_freqpoly()` have a binwidth argument; `geom_density()` has a `bw` argument.

3. Modify the app you created in the previous exercise to allow the user to choose whether each geom is shown or not (i.e., instead of always using one geom, they can pick 0, 1, 2, or 3). Make sure that you can control the binwidth of the histogram and frequency polygon independently.

Creating UI with Code

Sometimes none of the techniques previously described gives you the level of dynamism that you need: the update functions only allow you to change existing inputs, and a tabset only works if you have a fixed and known set of possible combinations. Sometimes you need to create different types or numbers of inputs (or outputs), depending on other inputs. This final technique gives you the ability to do so.

It's worth noting that you've always created your user interface with code, but so far you've always done it before the app starts. This technique gives you the ability to create and modify the user interface while the app is running. There are two parts to this solution:

- `uiOutput()` inserts a placeholder in your `ui`. This leaves a "hole" that your server code can later fill in.

- `renderUI()` is called within `server()` to fill in the placeholder with dynamically generated UI.

We'll see how this works with a simple example and then dive into some realistic uses.

Getting Started

Let's begin with a simple app that dynamically creates an input control, with the type and label control by two other inputs:

```
ui <- fluidPage(
  textInput("label", "label"),
  selectInput("type", "type", c("slider", "numeric")),
  uiOutput("numeric")
)
server <- function(input, output, session) {
  output$numeric <- renderUI({
    if (input$type == "slider") {
      sliderInput("dynamic", input$label, value = 0, min = 0, max = 10)
    } else {
      numericInput("dynamic", input$label, value = 0, min = 0, max = 10)
    }
  })
}
```

The resulting app is shown in Figure 10-8.

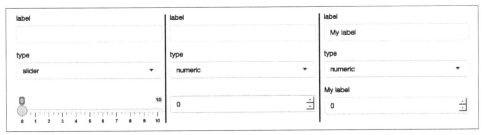

Figure 10-8. App on load (left), then changing type to numeric (middle), then label to "my label." See live at https://hadley.shinyapps.io/ms-render-simple.

If you run this code yourself, you'll notice that it takes a fraction of a second to appear after the app loads. That's because it's reactive: the app must load and trigger a reactive event, which calls the server function, yielding HTML to insert into the page. This is one of the downsides of renderUI(); relying on it too much can create a laggy UI. For good performance, strive to keep fixed as much of the user interface as possible, using the techniques described earlier in the chapter.

There's one other problem with this approach: when you change controls, you lose the currently selected value. Maintaining existing state is one of the big challenges of creating UI with code. This is one reason that selectively showing and hiding UI is a better approach if it works for you—because you're not destroying and re-creating the controls, you don't need to do anything to preserve the values. However, in many cases, we can fix the problem by setting the value of the new input to the current value of the existing control:

```
server <- function(input, output, session) {
  output$numeric <- renderUI({
    value <- isolate(input$dynamic)
    if (input$type == "slider") {
      sliderInput("dynamic", input$label, value = value, min = 0, max = 10)
    } else {
      numericInput("dynamic", input$label, value = value, min = 0, max = 10)
    }
  })
}
```

The use of isolate() is important. We'll come back to what it does in "isolate()" on page 228, but here it ensures that we don't create a reactive dependency that would cause this code to rerun every time input$dynamic changes (which will happen whenever the user modifies the value). We only want it to change when input$type or input$label changes.

Multiple Controls

Dynamic UI is most useful when you are generating an arbitrary number or type of controls. That means that you'll be generating UI with code, and I recommend using functional programming for this sort of task. Here I'll use purrr::map() and purrr::reduce(), but you could certainly do the same with the base lapply() and Reduce() functions:

```
library(purrr)
```

If you're not familiar with the map() and reduce() of functional programming, you might want to take a brief detour to read *Functional Programming* (*https://oreil.ly/mVxlM*) before continuing. We'll also come back to this idea in Chapter 18. These are complex ideas, so don't stress out if it doesn't make sense on your first read-through.

To make this concrete, imagine that you'd like the user to be able to supply their own color palette. They'll first specify how many colors they want and then supply a value for each color. The ui is pretty simple: we have a numericInput() that controls the number of inputs, a uiOutput() where the generated text boxes will go, and a textOutput() that demonstrates that we've plumbed everything together correctly:

```r
ui <- fluidPage(
  numericInput("n", "Number of colours", value = 5, min = 1),
  uiOutput("col"),
  textOutput("palette")
)
```

The server function is short but contains some big ideas:

```r
server <- function(input, output, session) {
  col_names <- reactive(paste0("col", seq_len(input$n)))

  output$col <- renderUI({
    map(col_names(), ~ textInput(.x, NULL))
  })

  output$palette <- renderText({
    map_chr(col_names(), ~ input[[.x]] %||% "")
  })
}
```

- I use a reactive, col_names(), to store the names of each of the color inputs I'm about to generate.

- I then use map() to create a list of textInput()s, one each for each name in col_names(). renderUI() then takes this list of HTML components and adds it to UI.

- I need to use a new trick to access the values the input values. So far we've always accessed the components of input with $, e.g., input$col1. But here we have the input names in a character vector, like var <- "col1". $ no longer works in this scenario, so we need to switch to [[, i.e., input[[var]].

- I use map_chr() to collect all values into a character vector and display that in output$palette. Unfortunately there's a brief period, just before the new inputs are rendered by the browser, where their values are NULL. This causes map_chr() to error, which we fix by using the handy %||% function: it returns the right-hand side whenever the left-hand side is NULL.

You can see the results in Figure 10-9.

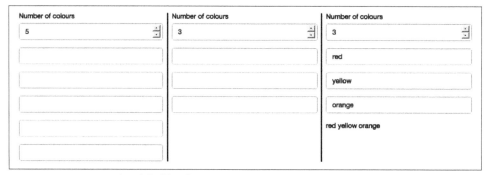

Figure 10-9. App on load (left), after setting n to 3 (middle), then entering some colors (right). See live at https://hadley.shinyapps.io/ms-render-palette.

If you run this app, you'll discover a really annoying behavior: whenever you change the number of colors, all the data you've entered disappears. We can fix this problem by using the same technique as before: setting `value` to the (isolated) current value. I'll also tweak the appearance to look a little nicer, including displaying the selected colors in a plot. Sample screenshots are shown in Figure 10-10:

```r
ui <- fluidPage(
  sidebarLayout(
    sidebarPanel(
      numericInput("n", "Number of colours", value = 5, min = 1),
      uiOutput("col"),
    ),
    mainPanel(
      plotOutput("plot")
    )
  )
)

server <- function(input, output, session) {
  col_names <- reactive(paste0("col", seq_len(input$n)))

  output$col <- renderUI({
    map(col_names(), ~ textInput(.x, NULL, value = isolate(input[[.x]])))
  })

  output$plot <- renderPlot({
    cols <- map_chr(col_names(), ~ input[[.x]] %||% "")
    # convert empty inputs to transparent
    cols[cols == ""] <- NA

    barplot(
      rep(1, length(cols)),
      col = cols,
      space = 0,
      axes = FALSE
    )
  })
}
```

```
    )
  }, res = 96)
}
```

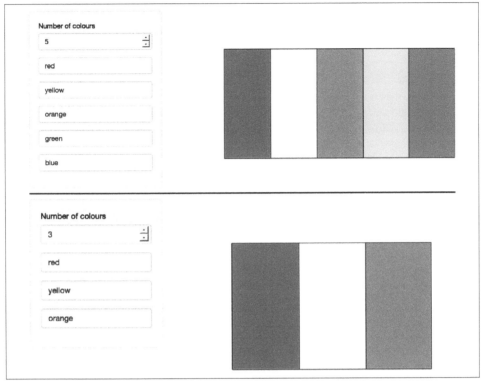

Figure 10-10. Filling out the colors of the rainbow (top), then reducing the number of colors to 3 (bottom); note that the existing colors are preserved. See live at https:// hadley.shinyapps.io/ms-render-palette-full.

Dynamic Filtering

To finish off the chapter, I'm going to create an app that lets you dynamically filter any data frame. Each numeric variable will get a range slider, and each factor variable will get a multiselect, so (for example) if a data frame has three numeric variables and two factors, the app will have three sliders and two select boxes.

I'll start with a function that creates the UI for a single variable. It'll return a range slider for numeric inputs, a multiselect for factor inputs, and NULL (nothing) for all other types:

```
make_ui <- function(x, var) {
  if (is.numeric(x)) {
    rng <- range(x, na.rm = TRUE)
    sliderInput(var, var, min = rng[1], max = rng[2], value = rng)
```

```
  } else if (is.factor(x)) {
    levs <- levels(x)
    selectInput(var, var, choices = levs, selected = levs, multiple = TRUE)
  } else {
    # Not supported
    NULL
  }
}
```

Then I'll write the server side equivalent of this function: it takes the variable and value of the input control and returns a logical vector saying whether or not to include each observation. Using a logical vector makes it easy to combine the results from multiple columns:

```
filter_var <- function(x, val) {
  if (is.numeric(x)) {
    !is.na(x) & x >= val[1] & x <= val[2]
  } else if (is.factor(x)) {
    x %in% val
  } else {
    # No control, so don't filter
    TRUE
  }
}
```

I can then use these functions "by hand" to generate a simple filtering UI for the iris dataset:

```
ui <- fluidPage(
  sidebarLayout(
    sidebarPanel(
      make_ui(iris$Sepal.Length, "Sepal.Length"),
      make_ui(iris$Sepal.Width, "Sepal.Width"),
      make_ui(iris$Species, "Species")
    ),
    mainPanel(
      tableOutput("data")
    )
  )
)
server <- function(input, output, session) {
  selected <- reactive({
    filter_var(iris$Sepal.Length, input$Sepal.Length) &
      filter_var(iris$Sepal.Width, input$Sepal.Width) &
      filter_var(iris$Species, input$Species)
  })

  output$data <- renderTable(head(iris[selected(), ], 12))
}
```

Figure 10-11 shows you the filter interface.

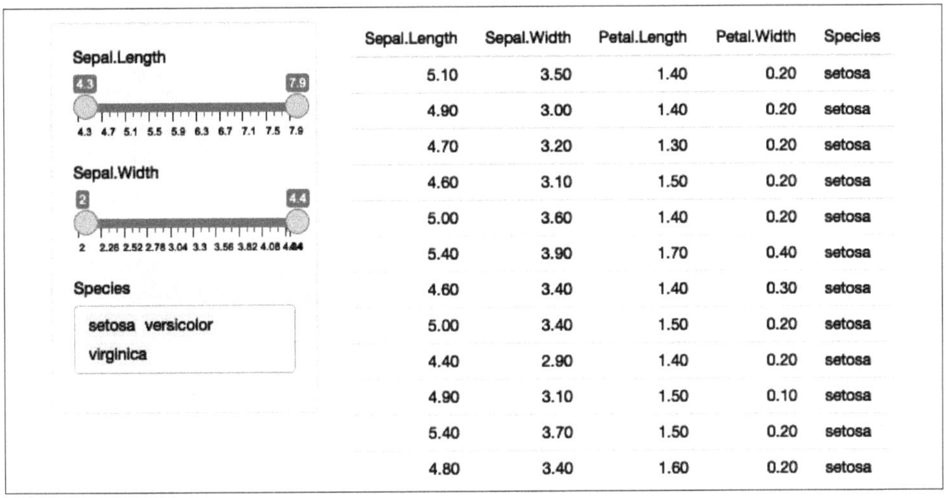
Sepal.Length
4.3
7.9
4.3 4.7 5.1 5.5 5.9 6.3 6.7 7.1 7.5 7.9
Sepal.Width
2
4.4
Species
setosa versicolor
virginica
Sepal.Length Sepal.Width Petal.Length Petal.Width Species
5.10 3.50 1.40 0.20 setosa
4.90 3.00 1.40 0.20 setosa
4.70 3.20 1.30 0.20 setosa
4.60 3.10 1.50 0.20 setosa
5.00 3.60 1.40 0.20 setosa
5.40 3.90 1.70 0.40 setosa
4.60 3.40 1.40 0.30 setosa
5.00 3.40 1.50 0.20 setosa
4.40 2.90 1.40 0.20 setosa
4.90 3.10 1.50 0.10 setosa
5.40 3.70 1.50 0.20 setosa
4.80 3.40 1.60 0.20 setosa

Figure 10-11. Simple filter interface for the iris dataset.

You might notice that I got sick of copying and pasting, so the app only works with three columns. I can make it work with all the columns by using a little functional programming:

- In `ui`, I use `map()` to generate one control for each variable.
- In `server()`, I use `map()` to generate the selection vector for each variable. Then I use `reduce()` to take the logical vector for each variable and combine into a single logical vector by `&`-ing each vector together.

Again, don't worry too much if you don't understand exactly what's happening here. The main takeaway is that once you master functional programming, you can write very succinct code that generates complex, dynamic apps:

```
ui <- fluidPage(
  sidebarLayout(
    sidebarPanel(
      map(names(iris), ~ make_ui(iris[[.x]], .x))
    ),
    mainPanel(
      tableOutput("data")
    )
  )
)
server <- function(input, output, session) {
  selected <- reactive({
    each_var <- map(names(iris), ~ filter_var(iris[[.x]], input[[.x]]))
    reduce(each_var, ~ .x & .y)
```

```
})

    output$data <- renderTable(head(iris[selected(), ], 12))
}
```

This creates what you see in Figure 10-12.

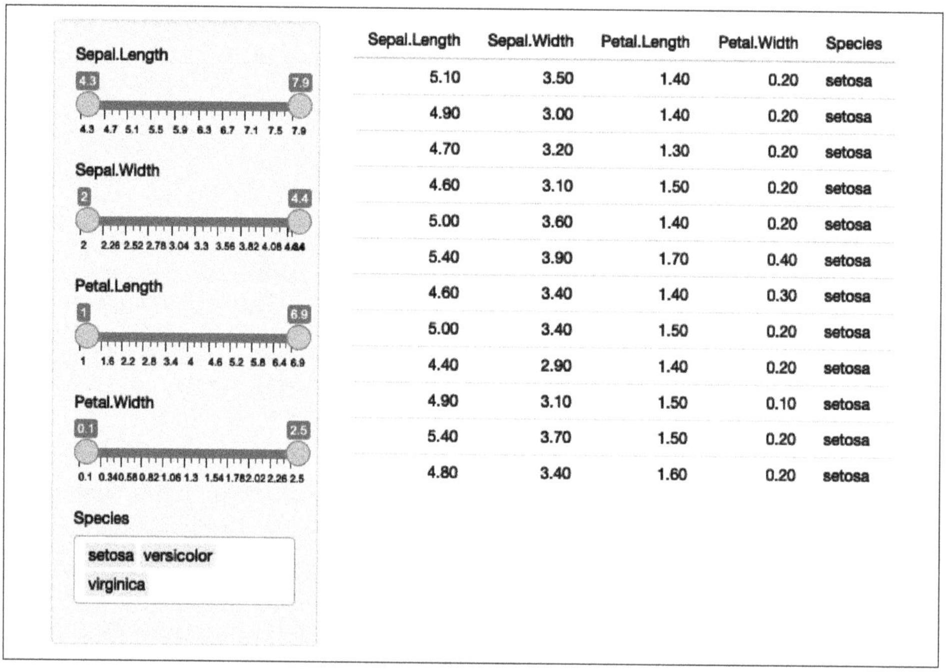

Sepal.Length	Sepal.Width	Petal.Length	Petal.Width	Species
5.10	3.50	1.40	0.20	setosa
4.90	3.00	1.40	0.20	setosa
4.70	3.20	1.30	0.20	setosa
4.60	3.10	1.50	0.20	setosa
5.00	3.60	1.40	0.20	setosa
5.40	3.90	1.70	0.40	setosa
4.60	3.40	1.40	0.30	setosa
5.00	3.40	1.50	0.20	setosa
4.40	2.90	1.40	0.20	setosa
4.90	3.10	1.50	0.10	setosa
5.40	3.70	1.50	0.20	setosa
4.80	3.40	1.60	0.20	setosa

Figure 10-12. Using functional programming to build a filtering app for the `iris` dataset.

From there, it's a simple generalization to work with any data frame. Here, I'll illustrate it using the data frames in the datasets package, but you can easily imagine how you might extend this to user-uploaded data:

```
dfs <- keep(ls("package:datasets"), ~ is.data.frame(get(.x, "package:datasets")))

ui <- fluidPage(
  sidebarLayout(
    sidebarPanel(
      selectInput("dataset", label = "Dataset", choices = dfs),
      uiOutput("filter")
    ),
    mainPanel(
      tableOutput("data")
    )
  )
)
server <- function(input, output, session) {
```

```
data <- reactive({
  get(input$dataset, "package:datasets")
})
vars <- reactive(names(data()))

output$filter <- renderUI(
  map(vars(), ~ make_ui(data()[[.x]], .x))
)

selected <- reactive({
  each_var <- map(vars(), ~ filter_var(data()[[.x]], input[[.x]]))
  reduce(each_var, `&`)
})

output$data <- renderTable(head(data()[selected(), ], 12))
}
```

See the result in Figure 10-13.

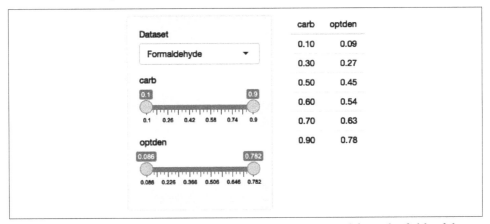

Figure 10-13. A dynamic user interface automatically generated from the fields of the selected dataset. See live at https://hadley.shinyapps.io/ms-filtering-final.

Dialog Boxes

Before we finish up, I wanted to mention a related technique: dialog boxes. You've seen them already in "Explicit Confirmation" on page 136, where the contents of the dialog was a fixed text string. But because `modalDialog()` is called from within the server function, you can actually dynamically generate content in the same way as `renderUI()`. This is a useful technique to have in your back pocket if you want to force the user to make some decision before continuing on with the regular app flow.

Exercises

1. Take this very simple app based on the initial example in the section:

```r
ui <- fluidPage(
  selectInput("type", "type", c("slider", "numeric")),
  uiOutput("numeric")
)
server <- function(input, output, session) {
  output$numeric <- renderUI({
    if (input$type == "slider") {
      sliderInput("n", "n", value = 0, min = 0, max = 100)
    } else {
      numericInput("n", "n", value = 0, min = 0, max = 100)
    }
  })
}
```

How could you instead implement it using dynamic visibility? If you implement dynamic visibility, how could you keep the values in sync when you change the controls?

2. Explain how this app works. Why does the password disappear when you click the "Enter password" button a second time?

```r
ui <- fluidPage(
  actionButton("go", "Enter password"),
  textOutput("text")
)
server <- function(input, output, session) {
  observeEvent(input$go, {
    showModal(modalDialog(
      passwordInput("password", NULL),
      title = "Please enter your password"
    ))
  })

  output$text <- renderText({
    if (!isTruthy(input$password)) {
      "No password"
    } else {
      "Password entered"
    }
  })
}
```

3. In the app in "Getting Started" on page 167, what happens if you drop the `iso late()` from `value <- isolate(input$dynamic)`?

4. Add support for date and date-time columns `make_ui()` and `filter_var()`.

5. (Advanced) If you know the S3 OOP (*http://adv-r.hadley.nz/S3.html*) system, consider how you could replace the `if` blocks in `make_ui()` and `filter_var()` with generic functions.

Summary

Before reading this chapter, you were limited to creating the user interface statically, before the server function was run. Now you've learned how to both modify the user interface and completely re-create it in response to user actions. A dynamic user interface will dramatically increase the complexity of your app, so don't be surprised if you find yourself struggling to debug what's going in. Always remember to use the simplest technique that solves your problem and fall back to the debugging advice in "Debugging" on page 70.

The next chapter switches tack to talk about bookmarking, which makes it possible to share the current state of an app with others.

## CHAPTER 11

Bookmarking

By default, Shiny apps have one major drawback compared to most websites: you can't bookmark the app to return to the same place in the future or share your work with someone else with a link in an email. That's because, by default, Shiny does not expose the current state of the app in its URL. Fortunately, however, you can change this behavior with a little extra work, and this chapter will show you how. As usual, we begin by loading shiny:

```
library(shiny)
```

Basic Idea

Let's take a simple app that we want to make bookmarkable. This app draws Lissajous figures, which replicate the motion of a pendulum. This app can produce a variety of interesting patterns that you might want to share:

```
ui <- fluidPage(
  sidebarLayout(
    sidebarPanel(
      sliderInput("omega", "omega", value = 1, min = -2, max = 2, step = 0.01),
      sliderInput("delta", "delta", value = 1, min = 0, max = 2, step = 0.01),
      sliderInput("damping", "damping", value = 1, min = 0.9, max = 1, step = 0.001),
      numericInput("length", "length", value = 100)
    ),
    mainPanel(
      plotOutput("fig")
    )
  )
)
server <- function(input, output, session) {
  t <- reactive(seq(0, input$length, length.out = input$length * 100))
  x <- reactive(sin(input$omega * t() + input$delta) * input$damping ^ t())
  y <- reactive(sin(t()) * input$damping ^ t())
```

```
output$fig <- renderPlot({
  plot(x(), y(), axes = FALSE, xlab = "", ylab = "", type = "l", lwd = 2)
}, res = 96)
}
```

Figure 11-1 shows the result.

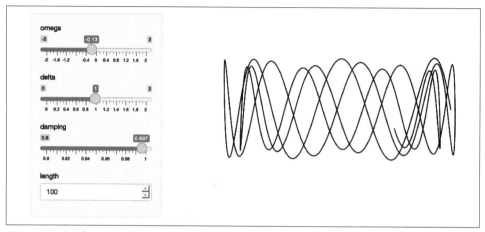

Figure 11-1. This app allows you to generate interesting figures using a model of a pendulum. Wouldn't it be cool to share a link with your friends?

There are three things we need to do to make this app bookmarkable:

1. Add a `bookmarkButton()` to the UI. This generates a button that the user clicks to generate the bookmarkable URL.

2. Turn `ui` into a function. You need to do this because bookmarked apps have to replay the bookmarked values: effectively, Shiny modifies the default `value` for each input control. This means there's no longer a single static UI but multiple possible UIs that depend on parameters in the URL (i.e., it has to be a function).

3. Add `enableBookmarking = "url"` to the `shinyApp()` call.

Making those changes gives us:

```
ui <- function(request) {
  fluidPage(
    sidebarLayout(
      sidebarPanel(
        sliderInput("omega", "omega", value = 1, min = -2, max = 2, step = 0.01),
        sliderInput("delta", "delta", value = 1, min = 0, max = 2, step = 0.01),
        sliderInput("damping", "damping", value = 1, min = 0.9, max = 1, step = 0.001),
        numericInput("length", "length", value = 100),
        bookmarkButton()
      ),
      mainPanel(
        plotOutput("fig")
```

```
          )
        )
      )
    }
shinyApp(ui, server, enableBookmarking = "url")
```

If you play around in the live app (*https://hadley.shinyapps.io/ms-bookmark-url*) and bookmark a few interesting states, you'll see that the generated URLs look something like this:

```
https://hadley.shinyapps.io/ms-bookmark-url/
  ?_inputs_&damping=1&delta=1&length=100&omega=1

https://hadley.shinyapps.io/ms-bookmark-url/
  ?_inputs_&damping=0.966&delta=1.25&length=100&omega=-0.54

https://hadley.shinyapps.io/ms-bookmark-url/
  ?_inputs_&damping=0.997&delta=1.37&length=500&omega=-0.9
```

To understand what's happening, let's take the first URL and tease it apart into pieces:

- `http://` is the "protocol" used to communicate with the app. This will always be `http` or `https`.

- `hadley.shinyapps.io/ms-bookmark-url` is the location of the app.

- Everything after ? is a "parameter." Each parameter is separated by &, and if you break it apart, you can see the values of each input in the app:

  — `damping=1`

  — `delta=1`

  — `length=100`

  — `omega=1`

So "generating a bookmark" means recording the current values of the inputs in the parameters of URL. If you play around with the app locally, the URLs will look slightly different:

```
http://127.0.0.1:4087/
  ?_inputs_&damping=1&delta=1&length=100&omega=1

http://127.0.0.1:4087/
  ?_inputs_&damping=0.966&delta=1.25&length=100&omega=-0.54

http://127.0.0.1:4087/
  ?_inputs_&damping=0.997&delta=1.37&length=500&omega=-0.9
```

Most of the pieces are the same except that instead of `hadley.shinyapps.io/ms-bookmark-url`, you'll see something like `127.0.0.1:4087`. `127.0.0.1` is a special address that always points to your own computer, and `4087` is a randomly assigned port. Normally, different apps get different paths or IP addresses, but that's not possible when you're hosting multiple apps on your own computer.

Updating the URL

Instead of providing an explicit button, another option is to automatically update the URL in the browser. This allows your users to use the user bookmark command in their browser or copy and paste the URL from the location bar.

Automatically updating the URL requires a little boilerplate in the server function:

```
# Automatically bookmark every time an input changes
observe({
  reactiveValuesToList(input)
  session$doBookmark()
})
# Update the query string
onBookmarked(updateQueryString)
```

This gives us an updated server function as follows:

```
server <- function(input, output, session) {
  t <- reactive(seq(0, input$length, length = input$length * 100))
  x <- reactive(sin(input$omega * t() + input$delta) * input$damping ^ t())
  y <- reactive(sin(t()) * input$damping ^ t())

  output$fig <- renderPlot({
    plot(x(), y(), axes = FALSE, xlab = "", ylab = "", type = "l", lwd = 2)
  }, res = 96)

  observe({
    reactiveValuesToList(input)
    session$doBookmark()
  })
  onBookmarked(updateQueryString)
}

shinyApp(ui, server, enableBookmarking = "url")
```

You can see what this yields in the live app (*https://hadley.shinyapps.io/ms-bookmark-auto*). Since the URL now automatically updates, you could now remove the bookmark button from the UI.

Storing Richer State

So far we've used `enableBookmarking = "url"`, which stores the state directly in the URL. This is a good place to start because it's very simple and works everywhere you might deploy your Shiny app. As you can imagine, however, the URL is going to get very long if you have a large number of inputs, and it's obviously not going to be able to capture an uploaded file.

For these cases, you might instead want to use `enableBookmarking = "server"`, which saves the state to an *.rds* file on the server. This always generates a short, opaque URL but requires additional storage on the server:

```
shinyApp(ui, server, enableBookmarking = "server")
```

shinyapps.io doesn't currently support server-side bookmarking, so you'll need to try this out locally. If you do so, you'll see that the bookmark button generates URLs like:

```
http://127.0.0.1:4087/?_state_id_=0d645f1b28f05c97

http://127.0.0.1:4087/?_state_id_=87b56383d8a1062c

http://127.0.0.1:4087/?_state_id_=c8b0291ba622b69c
```

These are paired with matching directories in your working directory:

```
shiny_bookmarks/0d645f1b28f05c97

shiny_bookmarks/87b56383d8a1062c

shiny_bookmarks/c8b0291ba622b69c
```

The main drawbacks with server bookmarking is that it requires files to be saved on the server, and it's not obvious how long these need to hang around for. If you're bookmarking complex state and you never delete these files, your app is going to take up more and more disk space over time. If you do delete the files, some old bookmarks are going to stop working.

Bookmarking Challenges

Automated bookmarking relies on the reactive graph. It seeds the inputs with the saved values, then replays all reactive expressions and outputs, which will yield the same app that you see, as long as your app's reactive graph is straightforward. This section briefly covers some of the cases, which need a little extra care:

- If your app uses random numbers, the results might be different even if all the inputs are the same. If it's really important to always generate the same numbers, you'll need to think about how to make your random process reproducible. The easiest way to do this is to use `repeatable()`; see the documentation for more details.

- If you have tabs and want to bookmark and restore the active tab, make sure to supply an `id` in your call to `tabsetPanel()`.

- If there are inputs that should not be bookmarked (e.g., they contain private information that shouldn't be shared), include a call to `setBookmarkExclude()` somewhere in your server function. For example,

  ```
  setBookmarkExclude(c("secret1", "secret2"))
  ```

will ensure that the `secret1` and `secret2` inputs are not bookmarked.

- If you are manually managing reactive state in your own `reactiveValues()` object (as we'll discuss in Chapter 16), you'll need to use the `onBookmark()` and `onRestore()` callbacks to manually save and load your additional state. See *Advanced Bookmarking* (*https://oreil.ly/S6D8c*) for more details.

Exercises

1. Generate an app for visualizing the results of ambient::noise_simplex() (*https://oreil.ly/UyK1G*). Your app should allow the user to control the frequency, fractal, lacunarity, and gain and be bookmarkable. How can you ensure the image looks exactly the same when reloaded from the bookmark? (Think about what the `seed` argument implies.)

2. Make a simple app that lets you upload a CSV file and then bookmark it. Upload a few files and then look in `shiny_bookmarks`. How do the files correspond to the bookmarks? (Hint: You can use `readRDS()` to look inside the cache files that Shiny is generating.)

Summary

This chapter has shown how to enable bookmarking for your app. This is a great feature to provide your users because it allows them to easily share their work with others. Next, we'll talk about how to use tidy evaluation within Shiny apps. Tidy evaluation is a feature of many tidyverse functions, and you'll need to learn about it if you want to allow the user to change variables in (for example) dplyr pipelines or ggplot2 graphics.

## CHAPTER 12

Tidy Evaluation

If you are using Shiny with the tidyverse, you will almost certainly encounter the challenge of programming with tidy evaluation. Tidy evaluation is used throughout the tidyverse to make interactive data exploration more fluid, but it comes with a cost: it's hard to refer to variables indirectly and hence harder to program with.

In this chapter, you'll learn how to wrap ggplot2 and dplyr functions in a Shiny app. (If you don't use the tidyverse, you can skip this chapter 😊.) The techniques for wrapping ggplot2 and dplyr functions in other functions or a package are a little different and are covered in other resources like *Using ggplot2 in Packages* (*https://oreil.ly/N0a1J*) or *Programming with dplyr* (*https://oreil.ly/4Mdfc*). Let's get started:

```
library(shiny)
library(dplyr, warn.conflicts = FALSE)
library(ggplot2)
```

Motivation

Imagine I want to create an app that allows you to filter a numeric variable to select rows that are greater than a threshold. You might write something like this:

```
num_vars <- c("carat", "depth", "table", "price", "x", "y", "z")
ui <- fluidPage(
  selectInput("var", "Variable", choices = num_vars),
  numericInput("min", "Minimum", value = 1),
  tableOutput("output")
)
server <- function(input, output, session) {
  data <- reactive(diamonds %>% filter(input$var > input$min))
  output$output <- renderTable(head(data()))
}
```

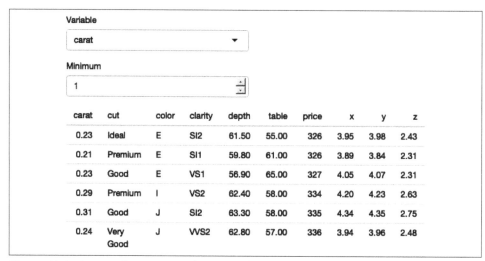

Variable									
carat							▼		

Minimum									
1									

carat	cut	color	clarity	depth	table	price	x	y	z
0.23	Ideal	E	SI2	61.50	55.00	326	3.95	3.98	2.43
0.21	Premium	E	SI1	59.80	61.00	326	3.89	3.84	2.31
0.23	Good	E	VS1	56.90	65.00	327	4.05	4.07	2.31
0.29	Premium	I	VS2	62.40	58.00	334	4.20	4.23	2.63
0.31	Good	J	SI2	63.30	58.00	335	4.34	4.35	2.75
0.24	Very Good	J	VVS2	62.80	57.00	336	3.94	3.96	2.48

Figure 12-1. An app that tries to select rows that are greater than a threshold on a user-selected variable.

As you can see from Figure 12-1, the app runs without error, but it doesn't return the correct result—all the rows have values of carat less than 1. The goal of the chapter is to help you understand why this doesn't work and why dplyr thinks you have asked for filter(diamonds, "carat" > 1).

This is a problem of *indirection*: normally when using tidyverse functions, you type the name of the variable directly in the function call. But now you want to refer to it indirectly: the variable (carat) is stored inside another variable (input$var).

That sentence might have made intuitive sense to you, but it's a bit confusing because I'm using "variable" to mean two slightly different things. It's going to be easier to understand what's happening if we can disambiguate the two uses by introducing two new terms:

env-variable
> An environment variable is a "programming" variable that you create with <-. input$var is an env-variable.

data-variable
> A data frame variable is a "statistical" variable that lives inside a data frame. carat is a data-variable.

With these new terms we can make the problem of indirection more clear: we have a data-variable (carat) stored inside an env-variable (input$var), and we need some way to tell dplyr this. There are two slightly different ways to do this, depending on whether the function you're working with is a "data-masking" function or a "tidy-selection" function.

Data-Masking

Data-masking functions allow you to use variables in the "current" data frame without any extra syntax. Data-masking is used in many dplyr functions like `arrange()`, `filter()`, `group_by()`, `mutate()`, and `summarise()`, and in ggplot2's `aes()`. Data-masking is useful because it lets you use data-variables without any additional syntax.

Getting Started

Let's begin with this call to `filter()`, which uses a data-variable (`carat`) and an env-variable (`min`):

```
min <- 1
diamonds %>% filter(carat > min)
#> # A tibble: 17,502 x 10
#>    carat cut        color clarity depth table price    x    y    z
#>    <dbl> <ord>      <ord> <ord>   <dbl> <dbl> <int> <dbl> <dbl> <dbl>
#> 1  1.17 Very Good  J     I1       60.2    61  2774  6.83  6.9  4.13
#> 2  1.01 Premium    F     I1       61.8    60  2781  6.39  6.36 3.94
#> 3  1.01 Fair       E     I1       64.5    58  2788  6.29  6.21 4.03
#> 4  1.01 Premium    H     SI2      62.7    59  2788  6.31  6.22 3.93
#> 5  1.05 Very Good  J     SI2      63.2    56  2789  6.49  6.45 4.09
#> 6  1.05 Fair       J     SI2      65.8    59  2789  6.41  6.27 4.18
#> # … with 17,496 more rows
```

Compare this to the base R equivalent:

```
diamonds[diamonds$carat > min, ]
```

In most, but not all,[1] base R functions, you have to refer to data-variables with $. This means that you often have to repeat the name of the data frame multiple times, but it does make it clear exactly what is a data-variable and what is an env-variable. It also makes it straightforward to use indirection[2] because you can store the name of the data-variable in an env-variable and then switch from $ to [[:

```
var <- "carat"
diamonds[diamonds[[var]] > min, ]
```

---

1 `dplyr::filter()` is inspired by `base::subset()`. `subset()` uses data-masking but not through tidy evaluation, so unfortunately the techniques discussed in this chapter don't apply to it.

2 In Shiny apps, the most common form of indirection is having the name of the data-variable stored in a reactive value. There's another form of indirection called *embracing* that happens when you're writing functions that is solved using {{ x }}. You can learn more about that in *Programming with dplyr* (*https://oreil.ly/4Mdfc*).

How can we achieve the same result with tidy evaluation? We need some way to add $ back into the picture. Fortunately, inside data-masking functions you can use `.data` or `.env` if you want to be explicit about whether you're talking about a data-variable or an env-variable:

```
diamonds %>% filter(.data$carat > .env$min)
```

Now we can switch from $ to [[:

```
diamonds %>% filter(.data[[var]] > .env$min)
```

Let's check that it works by updating the server function that we started the chapter with:

```
num_vars <- c("carat", "depth", "table", "price", "x", "y", "z")
ui <- fluidPage(
  selectInput("var", "Variable", choices = num_vars),
  numericInput("min", "Minimum", value = 1),
  tableOutput("output")
)
server <- function(input, output, session) {
  data <- reactive(diamonds %>% filter(.data[[input$var]] > .env$input$min))
  output$output <- renderTable(head(data()))
}
```

Figure 12-2 shows that we've been successful—we only see diamonds with values of `carat` greater than 1.

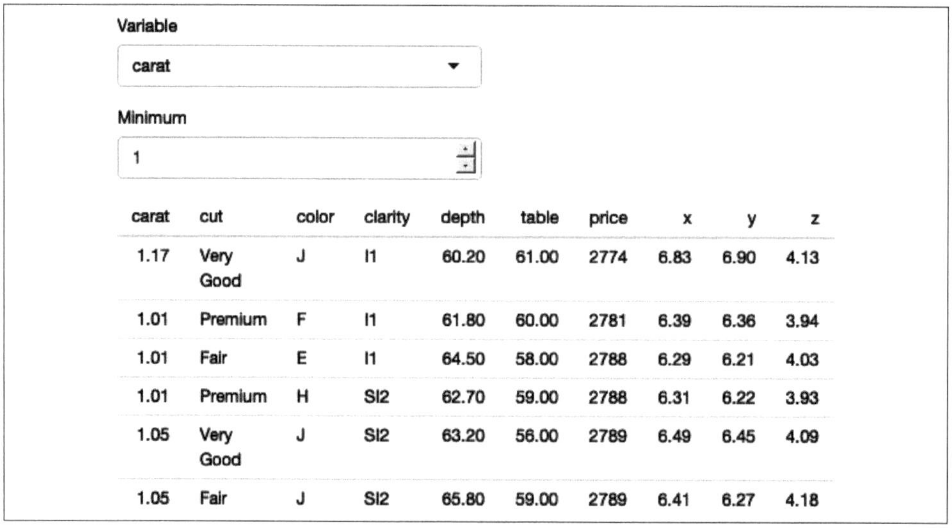

Variable
carat ▼

Minimum
1

carat	cut	color	clarity	depth	table	price	x	y	z
1.17	Very Good	J	I1	60.20	61.00	2774	6.83	6.90	4.13
1.01	Premium	F	I1	61.80	60.00	2781	6.39	6.36	3.94
1.01	Fair	E	I1	64.50	58.00	2788	6.29	6.21	4.03
1.01	Premium	H	SI2	62.70	59.00	2788	6.31	6.22	3.93
1.05	Very Good	J	SI2	63.20	56.00	2789	6.49	6.45	4.09
1.05	Fair	J	SI2	65.80	59.00	2789	6.41	6.27	4.18

Figure 12-2. Our app works now that we've been explicit about `.data` and `.env` and [[vs $. See live at https://hadley.shinyapps.io/ms-tidied-up.

Now that you've seen the basics, we'll develop a couple more realistic, but still simple, Shiny apps.

Example: ggplot2

Let's apply this idea to a dynamic plot where we allow the user to create a scatterplot by selecting the variables to appear on the x and y axes:

```
ui <- fluidPage(
  selectInput("x", "X variable", choices = names(iris)),
  selectInput("y", "Y variable", choices = names(iris)),
  plotOutput("plot")
)
server <- function(input, output, session) {
  output$plot <- renderPlot({
    ggplot(iris, aes(.data[[input$x]], .data[[input$y]])) +
      geom_point(position = ggforce::position_auto())
  }, res = 96)
}
```

The results are shown in Figure 12-3.

Here I've used `ggforce::position_auto()` so that `geom_point()` works nicely regardless of whether the x and y variables are continuous or discrete. Alternatively, we could allow the user to pick the geom. The following app uses a `switch()` statement to generate a reactive geom that is later added to the plot:

```
ui <- fluidPage(
  selectInput("x", "X variable", choices = names(iris)),
  selectInput("y", "Y variable", choices = names(iris)),
  selectInput("geom", "geom", c("point", "smooth", "jitter")),
  plotOutput("plot")
)
server <- function(input, output, session) {
  plot_geom <- reactive({
    switch(input$geom,
      point = geom_point(),
      smooth = geom_smooth(se = FALSE),
      jitter = geom_jitter()
    )
  })

  output$plot <- renderPlot({
    ggplot(iris, aes(.data[[input$x]], .data[[input$y]])) +
      plot_geom()
  }, res = 96)
}
```

This is one of the challenges of programming with user-selected variables: your code has to become more complicated to handle all the cases the user might generate.

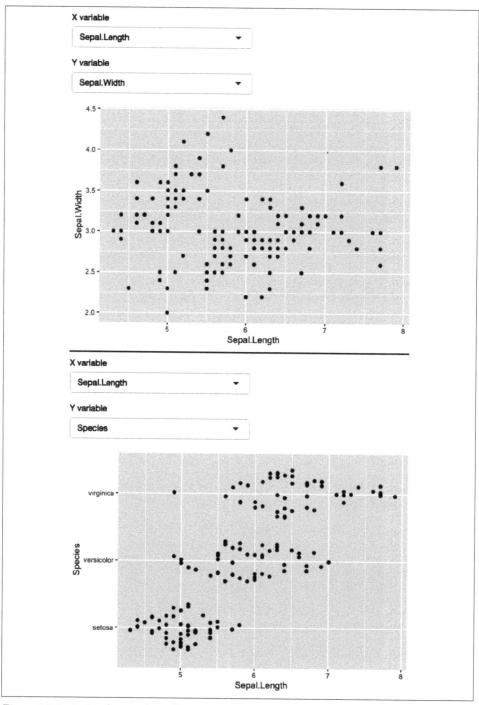

Figure 12-3. A simple app that allows you to select which variables are plotted on the x and y axes. See live at https://hadley.shinyapps.io/ms-ggplot2.

Example: dplyr

The same technique also works for dplyr. The following app extends the previous simple example to allow you to choose a variable to filter, a minimum value to select, and a variable to sort by:

```r
ui <- fluidPage(
  selectInput("var", "Select variable", choices = names(mtcars)),
  sliderInput("min", "Minimum value", 0, min = 0, max = 100),
  selectInput("sort", "Sort by", choices = names(mtcars)),
  tableOutput("data")
)
server <- function(input, output, session) {
  observeEvent(input$var, {
    rng <- range(mtcars[[input$var]])
    updateSliderInput(
      session, "min",
      value = rng[[1]],
      min = rng[[1]],
      max = rng[[2]]
    )
  })

  output$data <- renderTable({
    mtcars %>%
      filter(.data[[input$var]] > input$min) %>%
      arrange(.data[[input$sort]])
  })
}
```

Figure 12-4 shows the updated result.

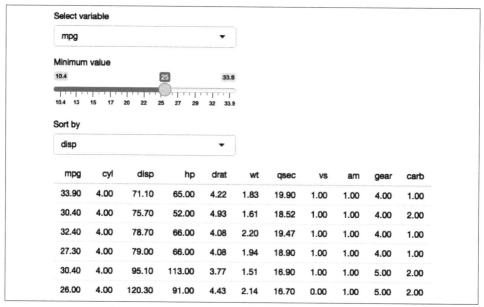

Figure 12-4. A simple app that allows you to pick a variable to threshold and choose how to sort the results. See live at https://hadley.shinyapps.io/ms-dplyr.

Most other problems can be solved by combining .data with your existing programming skills. For example, what if you wanted to conditionally sort in either ascending or descending order?

```r
ui <- fluidPage(
  selectInput("var", "Sort by", choices = names(mtcars)),
  checkboxInput("desc", "Descending order?"),
  tableOutput("data")
)
server <- function(input, output, session) {
  sorted <- reactive({
    if (input$desc) {
      arrange(mtcars, desc(.data[[input$var]]))
    } else {
      arrange(mtcars, .data[[input$var]])
    }
  })
  output$data <- renderTable(sorted())
}
```

As you provide more control, you'll find your code gets more and more complicated, and it becomes harder and harder to create a user interface that is both comprehensive and user-friendly. This is why I've always focused on code tools for data analysis: creating good UIs is really, really hard!

User-Supplied Data

Before we move on to talk about tidy-selection, there's one last topic we need to discuss: user-supplied data. Take this app shown in Figure 12-5: it allows the user to upload a TSV file, then select a variable and filter by it. It will work for the vast majority of inputs that you might try it with.

```r
ui <- fluidPage(
  fileInput("data", "dataset", accept = ".tsv"),
  selectInput("var", "var", character()),
  numericInput("min", "min", 1, min = 0, step = 1),
  tableOutput("output")
)
server <- function(input, output, session) {
  data <- reactive({
    req(input$data)
    vroom::vroom(input$data$datapath)
  })
  observeEvent(data(), {
    updateSelectInput(session, "var", choices = names(data()))
  })
  observeEvent(input$var, {
    val <- data()[[input$var]]
    updateNumericInput(session, "min", value = min(val))
  })

  output$output <- renderTable({
    req(input$var)

    data() %>%
      filter(.data[[input$var]] > input$min) %>%
      arrange(.data[[input$var]]) %>%
      head(10)
  })
}
```

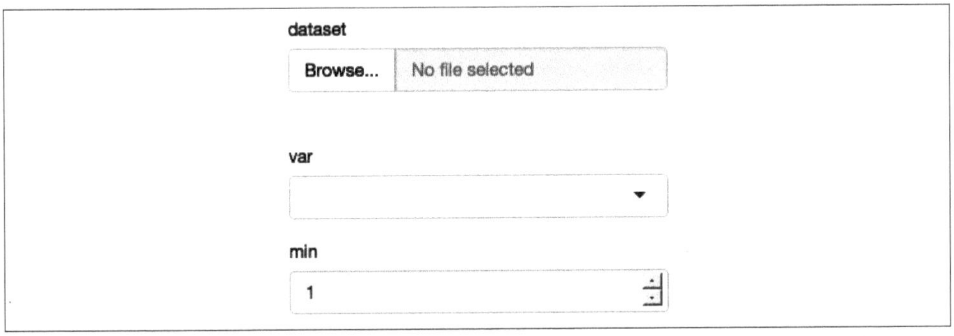

Figure 12-5. An app that filters user-supplied data, with a surprising failure mode. See live at https://hadley.shinyapps.io/ms-user-supplied.

There is a subtle problem with the use of `filter()` here. Let's pull out the call to `filter()` so we can play around with it directly, outside of the app:

```
df <- data.frame(x = 1, y = 2)
input <- list(var = "x", min = 0)

df %>% filter(.data[[input$var]] > input$min)
#>   x y
#> 1 1 2
```

If you experiment with this code, you'll find that it appears to work just fine for the vast majority of data frames. However, there's a subtle issue: What happens if the data frame contains a variable called `input`?

```
df <- data.frame(x = 1, y = 2, input = 3)
df %>% filter(.data[[input$var]] > input$min)
#> Error: Problem with `filter()` input `..1`.
#> x $ operator is invalid for atomic vectors
#> i Input `..1` is `.data[["x"]] > input$min`.
```

We get an error message because `filter()` is attempting to evaluate `df$input$min`:

```
df$input$min
#> Error in df$input$min: $ operator is invalid for atomic vectors
```

This problem is due to the ambiguity of data-variables and env-variables and because data-masking prefers to use a data-variable if both are available. We can resolve the problem by using `.env`[3] to tell `filter()` to only look for `min` in the env-variables:

```
df %>% filter(.data[[input$var]] > .env$input$min)
#>   x y input
#> 1 1 2     3
```

Note that you only need to worry about this problem when working with user-supplied data; when working with your own data, you can ensure the names of your data-variables don't clash with the names of your env-variables (and if they accidentally do, you'll discover it right away).

Why Not Use Base R?

At this point you might wonder if you're better off without `filter()` and if instead you should use the equivalent base R code:

```
df[df[[input$var]] > input$min, ]
#>   x y input
#> 1 1 2     3
```

---

3 You might wonder if the same problem applies to variables called `.data` and `.env`. In the unlikely event of having columns with those names, you'll need to refer to them explicitly with `.data$.data` and `.data$.env`.

That's a totally legitimate position, as long as you're aware of the work that `filter()` does for you so you can generate the equivalent base R code. In this case:

- You'll need `drop` = `FALSE` if `df` only contains a single column (otherwise you'll get a vector instead of a data frame).
- You'll need to use `which()` or similar to drop any missing values.
- You can't do group-wise filtering (e.g., `df %>% group_by(g) %>% filter(n() == 1)`).

In general, if you're using dplyr for very simple cases, you might find it easier to use base R functions that don't use data-masking. However, in my opinion, one of the advantages of the tidyverse is the careful thought that has been applied to edge cases so that functions work more consistently. I don't want to oversell this, but at the same time, it's easy to forget the quirks of specific base R functions and write code that works 95% of the time but fails in unusual ways the other 5% of the time.

Tidy-Selection

As well as data-masking, there's one other important part of tidy evaluation: tidy-selection. Tidy-selection provides a concise way of selecting columns by position, name, or type. It's used in `dplyr::select()` and `dplyr::across()` and in many functions from tidyr, like `pivot_longer()`, `pivot_wider()`, `separate()`, `extract()`, and `unite()`.

Indirection

To refer to variables indirectly, use `any_of()` or `all_of()`:[4] both expect a character vector env-variable containing the names of data-variables. The only difference is what happens if you supply a variable name that doesn't exist in the input: `all_of()` will throw an error, while `any_of()` will silently ignore it.

For example, the following app lets the user select any number of variables using a multiselect input, along with `all_of()`:

```
ui <- fluidPage(
  selectInput("vars", "Variables", names(mtcars), multiple = TRUE),
  tableOutput("data")
)

server <- function(input, output, session) {
```

---

4 In older versions of tidyselect and dplyr, you'll need to use `one_of()`. It has the same semantics as `any_of()` but a less informative name.

```
output$data <- renderTable({
  req(input$vars)
  mtcars %>% select(all_of(input$vars))
})
}
```

Tidy-Selection and Data-Masking

Working with multiple variables is trivial when you're working with a function that
uses tidy-selection: you can just pass a character vector of variable names into
any_of() or all_of(). Wouldn't it be nice if we could do that in data-masking functions too? That's the idea of the across() function, added in dplyr 1.0.0. It allows you
to use tidy-selection inside data-masking functions.

across() is typically used with either one or two arguments. The first argument
selects variables and is useful in functions like group_by() or distinct(). For example, the app in Figure 12-4 allows you to select any number of variables and count
their unique values:

```
ui <- fluidPage(
  selectInput("vars", "Variables", names(mtcars), multiple = TRUE),
  tableOutput("count")
)

server <- function(input, output, session) {
  output$count <- renderTable({
    req(input$vars)

    mtcars %>%
      group_by(across(all_of(input$vars))) %>%
      summarise(n = n(), .groups = "drop")
  })
}
```

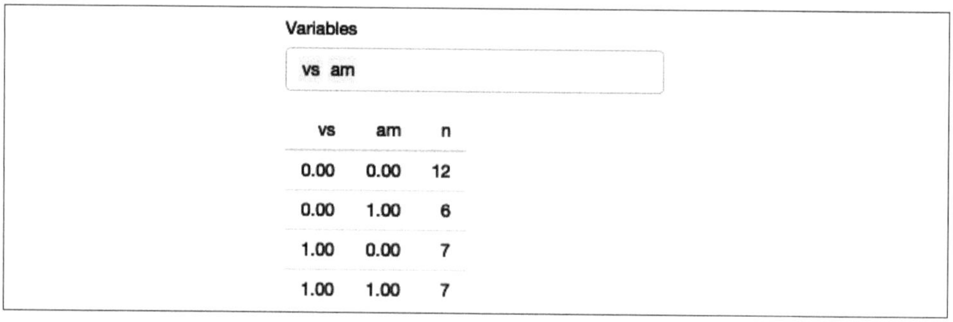

*Figure 12-6. This app allows you to select any number of variables and count their
unique combinations. See live at https://hadley.shinyapps.io/ms-across.*

The second argument is a function (or list of functions) that's applied to each selected column. That makes it a good fit for `mutate()` and `summarise()` where you typically want to transform each variable in some way. For example, the following code lets the user select any number of grouping variables and any number of variables to summarize with their means:

```
ui <- fluidPage(
  selectInput("vars_g", "Group by", names(mtcars), multiple = TRUE),
  selectInput("vars_s", "Summarise", names(mtcars), multiple = TRUE),
  tableOutput("data")
)

server <- function(input, output, session) {
  output$data <- renderTable({
    mtcars %>%
      group_by(across(all_of(input$vars_g))) %>%
      summarise(across(all_of(input$vars_s), mean), n = n())
  })
}
```

parse() and eval()

Before we conclude, it's worth a brief comment about `paste()` + `parse()` + `eval()`. If you have no idea what this combination is, you can skip this section, but if you have used it, I'd like to pass on a small note of caution.

It's a tempting approach because it requires learning very few new ideas. But it has some major downsides: because you are pasting strings together, it's very easy to accidentally create invalid code or code that can be abused to do something that you didn't want. This isn't super important if it's a Shiny app that only you use, but it isn't a good habit to get into—otherwise it's very easy to accidentally create a security hole in an app that you share more widely. We'll come back that idea in Chapter 22.

(You shouldn't feel bad if this is the only way you can figure out to solve a problem, but when you have a bit more mental space, I'd recommend spending some time figuring out how to do it without string manipulation. This will help you to become a better R programmer.)

Summary

In this chapter you've learned how to create Shiny apps that let the user choose which variables will be fed into tidyverse functions like `dplyr::filter()` and `ggplot2::aes()`. This requires getting your head around a key distinction that you haven't had to think about before: the difference between a data-variable and an env-variable. It will take some practice before this becomes second nature, but once you master the ideas, you unlock the power to expose the data analysis powers of the tidy-verse to non-R users.

This is the last chapter in the "Shiny in Action" part of the book. Now that you have the tools you need to make a range of useful apps, I'm going to focus on improving your understanding of the theory that underlies Shiny.

## PART III

Mastering Reactivity

You now have a bundle of useful techniques under your belt, giving you the ability to create a wide range of useful apps. Next we'll turn our attention to the theory of reactivity that underlies the magic of Shiny:

- In Chapter 13, you'll learn why the reactivity programming model is needed and a little bit about the history of reactive programming outside of R.
- In Chapter 14, you'll learn the full details of the reactive graph, which determines exactly when reactive components are updated.
- In Chapter 15, you'll learn about the underlying building blocks, particularly observers and timed invalidation.
- In Chapter 16, you'll learn how to escape the constraints of the reactive graph using `reactiveVal()` and `observe()`.

You certainly don't need to understand all these details for routine development of Shiny apps. But improving your understanding will help you write correct apps from the get-go, and when something behaves unexpectedly, you can more quickly narrow in on the underlying issue.

## CHAPTER 13

Why Reactivity?

Introduction

The initial impression of Shiny is often that it's "magic." Magic is great when you get started because you can make simple apps very, very quickly. But magic in software usually leads to disillusionment: without a solid mental model, it's extremely difficult to predict how the software will act when you venture beyond the borders of its demos and examples. And when things don't go the way you expect, debugging is almost impossible.

Fortunately, Shiny is "good" magic. As Tom Dale said of his Ember.js JavaScript framework: "We do a lot of magic, but it's *good magic*, which means it decomposes into sane primitives."[1] This is the quality that the Shiny team aspires to for Shiny, especially when it comes to reactive programming. When you peel back the layers of reactive programming, you won't find a pile of heuristics, special cases, and hacks; instead you'll find a clever but ultimately fairly straightforward mechanism. Once you've formed an accurate mental model of reactivity, you'll see that there's nothing up Shiny's sleeves: the magic comes from simple concepts combined in consistent ways.

In this chapter, we'll motivate reactive programming by trying to do without it and then give a brief history of reactivity as it pertains to Shiny.

---

[1] Steve Sanderson, "Rich JavaScript Applications—The Seven Frameworks (Throne of JS, 2012)," (*https://oreil.ly/HBSKI*) Steve Sanderson's Blog, August 1, 2012.

Why Do We Need Reactive Programming?

Reactive programming is a style of programming that focuses on values that change over time and calculations and actions that depend on those values. Reactivity is important for Shiny apps because they're interactive: users change input controls (dragging sliders, typing in text boxes, checking checkboxes, etc.), which causes logic to run on the server (reading CSVs, subsetting data, fitting models, etc.), ultimately resulting in outputs updating (plots redrawing, tables updating, etc.). This is quite different from most R code, which typically deals with fairly static data.

For Shiny apps to be maximally useful, we need reactive expressions and outputs to update if and only if their inputs change. We want outputs to stay in sync with inputs while ensuring that we never do more work than necessary. To see why reactivity is so helpful here, we'll take a stab at solving a simple problem without reactivity.

Why Can't You Use Variables?

In one sense, you already know how to handle "values that change over time": they're called *variables*. Variables in R represent values, and they can change over time, but they're not designed to help you when they change. Take this simple example of converting a temperature from Celsius to Fahrenheit:

```
temp_c <- 10
temp_f <- (temp_c * 9 / 5) + 32
temp_f
#> [1] 50
```

So far, so good: the `temp_c` variable has the value 10, the `temp_f` variable has the value 50, and we can change `temp_c`:

```
temp_c <- 30
```

But changing `temp_c` does not affect `temp_f`:

```
temp_f
#> [1] 50
```

Variables can change over time, but they never change automatically.

What About Functions?

You could instead attack this problem with a function:

```
temp_c <- 10
temp_f <- function() {
  message("Converting")
  (temp_c * 9 / 5) + 32
}
temp_f()
```

```
#> Converting
#> [1] 50
```

(This is a slightly weird function because it doesn't have any arguments, instead accessing temp_c from its enclosing environment,[2] but it's perfectly valid R code.)

This solves the first problem that reactivity is trying to solve: whenever you access temp_f(), you get the latest computation:

```
temp_c <- -3
temp_f()
#> Converting
#> [1] 26.6
```

It doesn't, however, minimize computation. Every time you call temp_f(), it recomputes, even if temp_c hasn't changed:

```
temp_f()
#> Converting
#> [1] 26.6
```

Computation is cheap in this trivial example, so needlessly repeating it isn't a big deal, but it's still unnecessary: if the inputs haven't changed, why do we need to recompute the output?

Event-Driven Programming

Since neither variables nor functions work, we need to create something new. In previous decades, we would've jumped directly to *event-driven programming*. Event-driven programming is an appealingly simple paradigm: you register callback functions that will be executed in response to events.

We could implement a very simple event-driven toolkit using R6, as in the following example. Here we define a DynamicValue that has three important methods: get() and set() to access and change the underlying value, and onUpdate() to register code to run whenever the value is modified. If you're not familiar with R6, don't worry about the details and instead focus on the following examples:

```
DynamicValue <- R6::R6Class("DynamicValue", list(
  value = NULL,
  on_update = NULL,

  get = function() self$value,

  set = function(value) {
    self$value <- value
    if (!is.null(self$on_update))
```

---

2 R uses "lexical scoping" (*https://oreil.ly/infBc*) for looking up the values associated with variable names.

```
      self$on_update(value)
    invisible(self)
  },

  onUpdate = function(on_update) {
    self$on_update <- on_update
    invisible(self)
  }
))
```

So if Shiny had been invented five years earlier, it might have looked more like this, where temp_c uses <<- to update temp_f whenever needed:[3]

```
temp_c <- DynamicValue$new()
temp_c$onUpdate(function(value) {
  message("Converting")
  temp_f <<- (value * 9 / 5) + 32
})

temp_c$set(10)
#> Converting
temp_f
#> [1] 50

temp_c$set(-3)
#> Converting
temp_f
#> [1] 26.6
```

Event-driven programming solves the problem of unnecessary computation, but it creates a new problem: you have to carefully track which inputs affect which computations. Before long, you start to trade off correctness (just update everything whenever anything changes) against performance (try to update only the necessary parts and hope that you didn't miss any edge cases) because it's so difficult to do both.

Reactive Programming

Reactive programming elegantly solves both problems by combining features of the preceding solutions. Now we can show you some real Shiny code, using a special Shiny mode, reactiveConsole(TRUE), that makes it possible to experiment with reactivity directly in the console:

```
library(shiny)
reactiveConsole(TRUE)
```

---

3 <<- is called the super-assignment operator (*https://oreil.ly/z26ra*), and here it modifies temp_f in the global environment rather than creating a new temp_f variable inside the function as <- would.

As with event-driven programming, we need some way to indicate that we have a special type of variable. In Shiny, we create a *reactive value* with reactiveVal(). A reactive value has special syntax[4] for getting its value (calling it like a zero-argument function) and setting its value (set its value by calling it like a one-argument function):

```
temp_c <- reactiveVal(10)  # create
temp_c()                   # get
#> [1] 10
temp_c(20)                 # set
temp_c()                   # get
#> [1] 20
```

Now we can create a reactive expression that depends on this value:

```
temp_f <- reactive({
  message("Converting")
  (temp_c() * 9 / 5) + 32
})
temp_f()
#> Converting
#> [1] 68
```

As you've learned when creating apps, a reactive expression automatically tracks all of its dependencies so that later, if temp_c changes, temp_f will automatically update:

```
temp_c(-3)
temp_c(-10)
temp_f()
#> Converting
#> [1] 14
```

But if temp_c() hasn't changed, then temp_f() doesn't need to recompute[5] and can just be retrieved from the cache:

```
temp_f()
#> [1] 14
```

A reactive expression has two important properties:

- It's *lazy*: it doesn't do any work until it's called.
- It's *cached*: it doesn't do any work the second and subsequent times it's called because it caches the previous result.

We'll come back to these important properties in Chapter 14.

---

4 If you happen to have ever used R's active bindings, you might notice that the syntax is very similar. This is not a coincidence.

5 You can tell it doesn't recompute because "Converting" is not printed.

A Brief History of Reactive Programming

If you want to learn more about reactive programming in other languages, a little history might be helpful. You can see the genesis of reactive programming over 40 years ago in VisiCalc (*https://oreil.ly/K4l08*), the first spreadsheet:

> I imagined a magic blackboard that if you erased one number and wrote a new thing in, all of the other numbers would automatically change, like word processing with numbers.
>
> —Dan Bricklin (*https://youtu.be/YDvbDiJZpy0*)

Spreadsheets are closely related to reactive programming: you declare the relationship between cells using formulas, and when one cell changes, all of its dependencies automatically update. So you've probably already done a bunch of reactive programming without knowing it!

While the ideas of reactivity have been around for a long time, it wasn't until the late 1990s that they were seriously studied in academic computer science. Research in reactive programming was kicked off by FRAN [@fran], functional reactive animation, a novel system for incorporating changes over time, and user input into a functional programming language. This spawned a rich literature [@rp-survey] but had little impact on the practice of programming.

It wasn't until the 2010s that reactive programming roared into the programming mainstream through the fast-paced world of JavaScript UI frameworks. Pioneering frameworks like Knockout (*https://knockoutjs.com*), Ember (*https://emberjs.com*), and Meteor (*https://www.meteor.com*) (Joe Cheng's personal inspiration for Shiny) demonstrated that reactive programming could make UI programming dramatically easier. Within a few short years, reactive programming has come to dominate web programming through hugely popular frameworks like React (*https://reactjs.org*), Vue.js (*https://vuejs.org*), and Angular (*https://angularjs.org*), which are all either inherently reactive or designed to work hand in hand with reactive backends.

It's worth bearing in mind that "reactive programming" is a fairly general term. While all reactive programming libraries, frameworks, and languages are broadly concerned with writing programs that respond to changing values, they vary enormously in their terminology, designs, and implementations. In this book, whenever we refer to "reactive programming," we are referring specifically to reactive programming as implemented in Shiny. So if you read material about reactive programming that isn't specifically about Shiny, it's unlikely that those concepts or even terminology will be relevant to writing Shiny apps. For readers who do have some experience with other reactive programming frameworks, Shiny's approach is similar to Meteor (*https://www.meteor.com*) and MobX (*https://mobx.js.org*) and very different from the ReactiveX (*http://reactivex.io*) family or anything that labels itself Functional Reactive Programming.

Summary

Now that you understand why reactive programming is needed and have learned a little bit of history, the next chapter will discuss more details of the underlying theory. Most importantly, you'll solidify your understanding of the reactive graph, which connects reactive values, reactive expressions, and observers and controls exactly what is run and when.

## CHAPTER 14

The Reactive Graph

Introduction

To understand reactive computation, you must first understand the reactive graph. In this chapter, we'll dive into the details of the graph, paying more attention to the precise order in which things happen. In particular, you'll learn about the importance of invalidation, the process that is key to ensuring that Shiny does the minimum amount of work. You'll also learn about the reactlog package, which can automatically draw the reactive graph for real apps.

If it's been a while since you looked at Chapter 3, I highly recommend that you refamiliarize yourself with it before continuing. It lays the groundwork for the concepts that we'll explore in more detail here.

A Step-by-Step Tour of Reactive Execution

To explain the process of reactive execution, we'll use the graphic shown in Figure 14-1. It contains three reactive inputs, three reactive expressions, and three outputs.[1] Recall that reactive inputs and expressions are collectively called reactive producers; reactive expressions and outputs are reactive consumers.

---

[1] Anywhere you see output, you can also think observer. The primary difference is that certain outputs that aren't visible will never be computed. We'll discuss the details in "Observers and Outputs" on page 226.

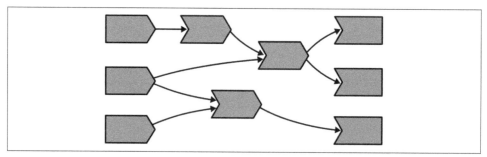

Figure 14-1. Complete reactive graph of an imaginary app containing three inputs, three reactive expressions, and three outputs.

The connections between the components are directional, with the arrows indicating the direction of reactivity. The direction might surprise you, as it's easy to think of a consumer taking dependencies on one or more producers. Shortly, however, you'll see that the flow of reactivity is more accurately modeled in the opposite direction.

The underlying app is not important, but if it helps you to have something concrete, you could pretend that it was derived from this not-very-useful app:

```r
ui <- fluidPage(
  numericInput("a", "a", value = 10),
  numericInput("b", "b", value = 1),
  numericInput("c", "c", value = 1),
  plotOutput("x"),
  tableOutput("y"),
  textOutput("z")
)

server <- function(input, output, session) {
  rng <- reactive(input$a * 2)
  smp <- reactive(sample(rng(), input$b, replace = TRUE))
  bc <- reactive(input$b * input$c)

  output$x <- renderPlot(hist(smp()))
  output$y <- renderTable(max(smp()))
  output$z <- renderText(bc())
}
```

Let's get started!

A Session Begins

Figure 14-2 shows the reactive graph right after the app has started and the server function has been executed for the first time.

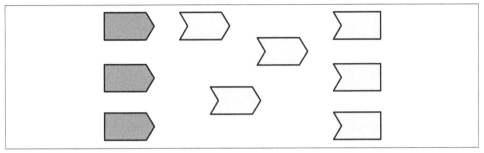

Figure 14-2. Initial state after app load. There are no connections between objects, and all reactive expressions are invalidated (gray). There are six reactive consumers and six reactive producers.

There are three important messages in this figure:

- There are no connections between the elements because Shiny has no a priori knowledge of the relationships between reactives.
- All reactive expressions and outputs are in their starting state, *invalidated* (gray), which means that they have yet to be run.
- The reactive inputs are ready (green), indicating that their values are available for computation.

Execution Begins

Now we start the execution phase, as shown in Figure 14-3.

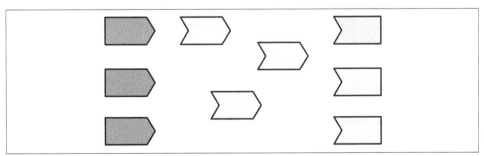

Figure 14-3. Shiny starts executing an arbitrary observer/output, colored yellow.

In this phase, Shiny picks an invalidated output and starts executing it (yellow). You might wonder how Shiny decides which of the invalidated outputs to execute. In short, you should act as if it's random: your observers and outputs shouldn't care what order they execute in, because they've been designed to function independently.[2]

Reading a Reactive Expression

Executing an output may require a value from a reactive, as in Figure 14-4.

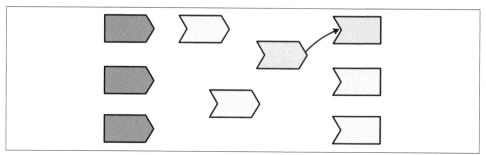

Figure 14-4. The output needs the value of a reactive expression, so it starts executing the expression.

Reading a reactive changes the graph in two ways:

- The reactive expression also needs to start computing its value (turn yellow). Note that the output is still computing: it's waiting on the reactive expression to return its value so its own execution can continue, just like a regular function call in R.

- Shiny records a relationship between the output and reactive expression (i.e., we draw an arrow). The direction of the arrow is important: the expression records that it is used by the output; the output doesn't record that it uses the expression. This is a subtle distinction, but its importance will become more clear when you learn about invalidation.

---

2 If you have observers whose side effects must happen in a certain order, you're generally better off redesigning your system. Failing that, you can control the relative order of observers with the priority argument to observe().

Reading an Input

This particular reactive expression happens to read a reactive input. Again, a dependency/dependent relationship is established, so in Figure 14-5 we add another arrow.

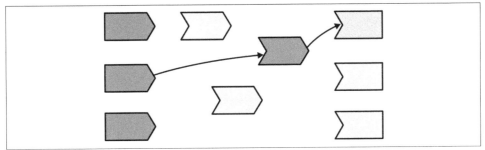

Figure 14-5. The reactive expression also reads from a reactive value, so we add another arrow.

Unlike reactive expressions and outputs, reactive inputs have nothing to execute so they can return immediately.

Reactive Expression Completes

In our example, the reactive expression reads another reactive expression, which in turn reads another input. We'll skip over the blow-by-blow description of those steps, since they're a repeat of what we've already described, and jump directly to Figure 14-6.

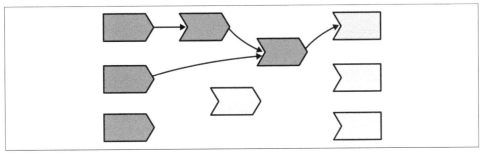

Figure 14-6. The reactive expression has finished computing, so it turns green.

Now that the reactive expression has finished executing, it turns green to indicate that it's ready. It caches the result so it doesn't need to recompute unless its inputs change.

Output Completes

Now that the reactive expression has returned its value, the output can finish executing and change color to green, as in Figure 14-7.

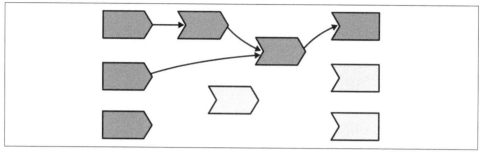

Figure 14-7. The output has finished computation and turns green.

The Next Output Executes

Now that the first output is complete, Shiny chooses another to execute. This output turns yellow, as seen in Figure 14-8, and starts reading values from reactive producers.

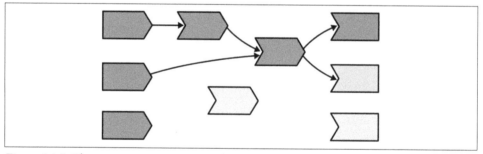

Figure 14-8. The next output starts computing, turning yellow.

Complete reactives can return their values immediately; invalidated reactives will kick off their own execution graph. This cycle will repeat until every invalidated output enters the complete (green) state.

Execution Completes, Outputs Flushed

Now all of the outputs have finished execution and are idle, as shown in Figure 14-9.

This round of reactive execution is complete, and no more work will occur until some external force acts on the system (e.g., the user of the Shiny app moving a slider in the user interface). In reactive terms, this session is now at rest.

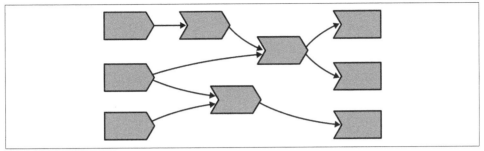

Figure 14-9. All output and reactive expressions have finished and turned green.

Let's stop here for a moment and think about what we've done. We've read some inputs, calculated some values, and generated some outputs. But more importantly, we also discovered the *relationships* between the reactive objects. When a reactive input changes, we know exactly which reactives we need to update.

An Input Changes

The previous step left off with our Shiny session in a fully idle state. Now imagine that the user of the application changes the value of a slider. This causes the browser to send a message to the server function, instructing Shiny to update the corresponding reactive input. This kicks off an *invalidation phase*, which has three parts: invalidating the input, notifying the dependencies, then removing the existing connections.

Invalidating the Inputs

The invalidation phase starts at the changed input/value, which we'll fill with gray, our usual color for invalidation, as in Figure 14-10.

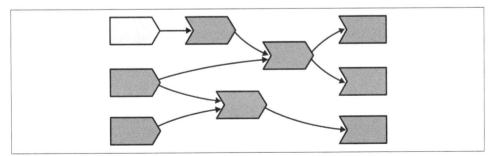

Figure 14-10. The user interacts with the app, invalidating an input.

Notifying Dependencies

Now, we follow the arrows that we drew earlier, coloring each node in gray and coloring the arrows we followed in light-gray. This yields Figure 14-11.

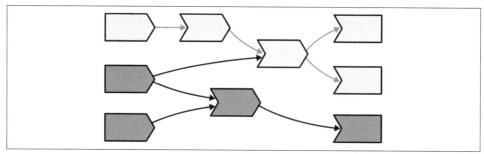

Figure 14-11. Invalidation flows out from the input, following every arrow from left to right. Arrows that Shiny has followed during invalidation are colored in a lighter gray.

Removing Relationships

Next, each invalidated reactive expression and output "erases" all of the arrows coming into and out of it, yielding Figure 14-12 and completing the invalidation phase.

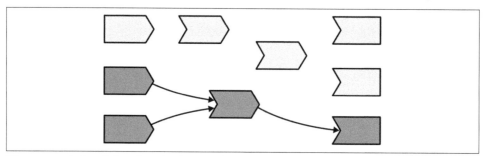

Figure 14-12. Invalidated nodes forget all their previous relationships so they can be discovered afresh.

The arrows coming out of a node are one-shot notifications that will fire the *next* time a value changes. Now that they've fired, they've fulfilled their purpose and we can erase them.

It's less obvious why we erase the arrows coming *into* an invalidated node, even if the node they're coming from isn't invalidated. While those arrows represent notifications that haven't fired, the invalidated node no longer cares about them: reactive consumers only care about notifications in order to invalidate themselves, and that that has already happened.

It may seem perverse that we put so much value on those relationships and now we've thrown them away! But this is a key part of Shiny's reactive programming model: though these particular arrows *were* important, they are now out of date. The only way to ensure that our graph stays accurate is to erase arrows when they become stale and let Shiny rediscover the relationships around these nodes as they re-execute. We'll come back to this important topic in "Dynamism" on page 218.

Re-execution

Now we're in a pretty similar situation to when we executed the second output, with a mix of valid and invalid reactives. It's time to do exactly what we did then: execute the invalidated outputs, one at a time, starting off in Figure 14-13.

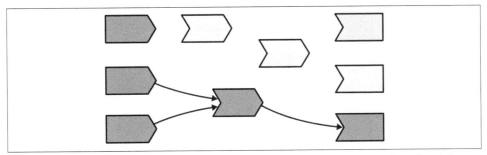

Figure 14-13. Now re-execution proceeds in the same way as execution, but there's less work to do since we're not starting from scratch.

Again, I won't show you the details, but the end result will be a reactive graph at rest, with all nodes marked in green. The neat thing about this process is that Shiny has done the minimum amount of work—we've only done the work needed to update the outputs that are actually affected by the changed inputs.

Exercises

1. Draw the reactive graph for the following server function and then explain why the reactives are not run:

   ```
   server <- function(input, output, session) {
     sum <- reactive(input$x + input$y + input$z)
     prod <- reactive(input$x * input$y * input$z)
     division <- reactive(prod() / sum())
   }
   ```

2. The following reactive graph simulates long-running computation by using `Sys.sleep()`:

   ```
   x1 <- reactiveVal(1)
   x2 <- reactiveVal(2)
   x3 <- reactiveVal(3)
   ```

```
y1 <- reactive({
  Sys.sleep(1)
  x1()
})
y2 <- reactive({
  Sys.sleep(1)
  x2()
})
y3 <- reactive({
  Sys.sleep(1)
  x2() + x3() + y2() + y2()
})

observe({
  print(y1())
  print(y2())
  print(y3())
})
```

How long will the graph take to recompute if x1 changes? What about x2 or x3?

3. What happens if you attempt to create a reactive graph with cycles?

```
x <- reactiveVal(1)
y <- reactive(x + y())
y()
```

Dynamism

In "Removing Relationships" on page 216, you learned that Shiny "forgets" the connections between reactive components that it spent so much effort recording. This makes Shiny's reactive dynamic because it can change while your app runs. This dynamism is so important that I want to reinforce it with a simple example:

```
ui <- fluidPage(
  selectInput("choice", "A or B?", c("a", "b")),
  numericInput("a", "a", 0),
  numericInput("b", "b", 10),
  textOutput("out")
)

server <- function(input, output, session) {
  output$out <- renderText({
    if (input$choice == "a") {
      input$a
    } else {
      input$b
    }
  })
}
```

You might expect the reactive graph to look like Figure 14-14.

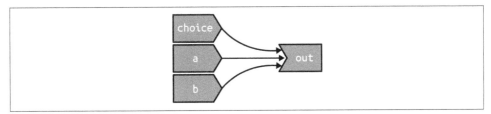

Figure 14-14. If Shiny analyzed reactivity statically, the reactive graph would always connect choice, a, and b to out.

But because Shiny dynamically reconstructs the graph after the output has been invalidated, it actually looks like either of the graphs in Figure 14-15, depending on the value of input$choice. This ensures that Shiny does the minimum amount of work when an input is invalidated. In this case, if input$choice is set to "b," then the value of input$a doesn't affect the output$out and there's no need to recompute it.

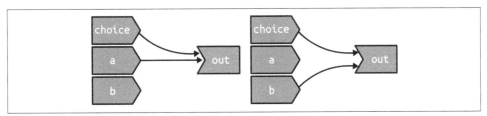

Figure 14-15. But Shiny's reactive graph is dynamic, so the graph either connects out to choice and a (left) or choice and b (right).

It's worth noting, as Yindeng Jiang does in their blog (*https://oreil.ly/6dDa4*), that a minor change will cause the output to always depend on both a and b:

```
output$out <- renderText({
  a <- input$a
  b <- input$b

  if (input$choice == "a") {
    a
  } else {
    b
  }
})
```

This would have no impact on the output of normal R code, but it makes a difference here because the reactive dependency is established when you read a value from input, not when you use that value.

The Reactlog Package

Drawing the reactive graph by hand is a powerful technique to help you understand simple apps and build up an accurate mental model of reactive programming. But it's painful to do for real apps that have many moving parts. Wouldn't it be great if we could automatically draw the graph using what Shiny knows about it? This is the job of the reactlog (*https://rstudio.github.io/reactlog*) package, which generates the so-called *reactlog*, which shows how the reactive graph evolves over time.

To see the reactlog, you'll need to first install the reactlog package, turn it on with `reactlog::reactlog_enable()`, then start your app. You then have two options:

- While the app is running, press Cmd+F3 (Ctrl+F3 on Windows) to show the reactlog generated up to that point.

- After the app has closed, run `shiny::reactlogShow()` to see the log for the complete session.

reactlog uses the same graphical conventions as this chapter, so it should feel instantly familiar. The biggest difference is that reactlog draws every dependency, even if it's not currently used, to keep the automated layout stable. Connections that are not currently active (but were in the past or will be in the future) are drawn as thin dotted lines.

Figure 14-16 shows the reactive graph that reactlog draws for the app we used earlier. There's a surprise in this screenshot: there are three additional reactive inputs (`client Data$output_x_height`, `clientData$output_x_width`, and `clientData$pixelra tio`) that don't appear in the source code. These exist because plots have an implicit dependency on the size of the output; whenever the output changes size, the plot needs to be redrawn.

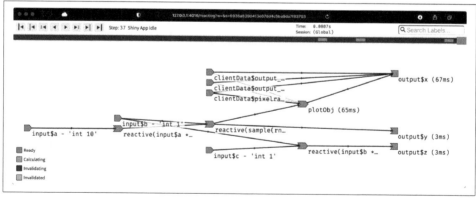

Figure 14-16. The reactive graph of our hypothetical app as drawn by reactlog.

Note that while reactive inputs and outputs have names, reactive expressions and observers do not, so they're labeled with their contents. To make things easier to understand, you may want use the `label` argument to `reactive()` and `observe()`, which will then appear in the reactlog. You can use emojis to make particularly important reactives stand out visually.

Summary

In this chapter, you've learned precisely how the reactive graph operates. In particular, you've learned for the first time about the invalidation phase, which doesn't immediately cause recomputation but instead marks reactive consumers as invalid so that they will be recomputed when needed. The invalidation cycle is also important because it clears out previously discovered dependencies so that they can be automatically rediscovered, making the reactive graphic dynamic.

Now that you've got the big picture under your belt, the next chapter will give some additional details about the underlying data structures that power reactive values, expressions, and output, and we'll discuss the related concept of timed invalidation.

## CHAPTER 15

Reactive Building Blocks

Now that you have the theory underpinning the reactive graph and you've got some practical experience, this is a good time to talk in more detail about how reactivity fits into R, the programming language. There are three fundamental building blocks of reactive programming: reactive values, reactive expressions, and observers. You've already seen most of the important parts of reactive values and expressions, so this chapter will spend more time on observers and outputs (which as you'll learn are a special type of observer). You'll also learn two other tools for controlling the reactive graph: isolation and timed invalidation.

This chapter will again use the reactive console so that we can experiment with reactivity directly in the console without having to launch a Shiny app each time. To begin, we'll load shiny and turn on reactivity for interactive experimentation:

```
library(shiny)
reactiveConsole(TRUE)
```

Reactive Values

There are two types of reactive values:

- A single reactive value, created by `reactiveVal()`
- A list of reactive values, created by `reactiveValues()`

They have slightly different interfaces for getting and setting values:

```
x <- reactiveVal(10)
x()       # get
#> [1] 10
x(20)     # set
x()       # get
```

```
#> [1] 20

r <- reactiveValues(x = 10)
r$x          # get
#> [1] 10
r$x <- 20 # set
r$x          # get
#> [1] 20
```

It's unfortunate that these two similar objects have rather different interfaces, but there's no way to standardize them. However, while they look different, they behave the same, so you can choose between them based on which syntax you prefer. In this book I use reactiveValues() because the syntax is easier to understand at a glance, but in my own code I tend to use reactiveVal() because the syntax makes it clear that something weird is going on.

It's important to note that both types of reactive values have so-called reference semantics. Most R objects have copy-on-modify (*https://oreil.ly/W1PzT*) semantics, which means that if you assign the same value to two names, the connection is broken as soon as you modify one:

```
a1 <- a2 <- 10
a2 <- 20
a1 # unchanged
#> [1] 10
```

This is not the case with reactive values—they always keep a reference back to the same value so that modifying any copy modifies all values:

```
b1 <- b2 <- reactiveValues(x = 10)
b1$x <- 20
b2$x
#> [1] 20
```

We'll come back to why you might create your own reactive values in Chapter 16. Otherwise, most of the reactive values you'll encounter will come from the input argument to the server function. These are a little different from the reactiveValues() that you create yourself because they're read-only: you can't modify the values because Shiny automatically updates them based on user actions in the browser.

Exercises

1. What are the differences between these two lists of reactive values? Compare the syntax for getting and setting individual reactive values:

   ```
   l1 <- reactiveValues(a = 1, b = 2)
   l2 <- list(a = reactiveVal(1), b = reactiveVal(2))
   ```

2. Design and perform a small experiment to verify that reactiveValue() also has reference semantics.

Reactive Expressions

Recall that a reactive has two important properties: it's lazy and cached. This means that it only does work when it's actually needed, and if called twice in a row, it returns the previous value.

There are two important details that we haven't yet covered: what reactive expressions do with errors and why on.exit() works inside of them.

Errors

Reactive expressions cache errors in exactly the same way that they cache values. For example, take this reactive:

```
r <- reactive(stop("Error occurred at ", Sys.time(), call. = FALSE))
r()
#> Error: Error occurred at 2021-03-05 16:38:19
```

If we wait a second or two, we can see that we get the same error as before:

```
Sys.sleep(2)
r()
#> Error: Error occurred at 2021-03-05 16:38:19
```

Errors are also treated the same way as values when it comes to the reactive graph: errors propagate through the reactive graph exactly the same way as regular values. The only difference is what happens when an error hits an output or observer:

- An error in an output will be displayed in the app.[1]
- An error in an observer will cause the current session to terminate. If you don't want this to happen, you'll need to wrap the code in try() or tryCatch().

This same system powers req() ("Canceling Execution with req()" on page 121), which emits a special type of error.[2] This special error causes observers and outputs to stop what they're doing but not to otherwise fail. By default, it will cause outputs to reset to their initial blank state, but if you use req(..., cancelOutput = TRUE), they'll preserve their current display.

---

1 By default, you'll see the whole error message. You can show a generic error message by turning error sanitizing (*https://oreil.ly/Q8HoR*) on.

2 Technically, a custom condition (*https://oreil.ly/kUHXg*).

on.exit()

You can think of `reactive(x())` as a shortcut for `function() x()`, automatically adding laziness and caching. This is mostly of importance if you want to understand how Shiny is implemented but means that you can use functions that only work inside functions. The most useful of these is `on.exit()`, which allows you to run code when a reactive expression finishes, regardless of whether the reactive successfully returns an error or fails with an error. This is what makes `on.exit()` work in "Removing on Completion" on page 128.

Exercises

1. Use the reactlog package to observe an error propagating through the reactives in the following app, confirming that it follows the same rules as value propagation:

   ```
   ui <- fluidPage(
     checkboxInput("error", "error?"),
     textOutput("result")
   )
   server <- function(input, output, session) {
     a <- reactive({
       if (input$error) {
         stop("Error!")
       } else {
         1
       }
     })
     b <- reactive(a() + 1)
     c <- reactive(b() + 1)
     output$result <- renderText(c())
   }
   ```

2. Modify the preceding app to use `req()` instead of `stop()`. Verify that events still propagate the same way. What happens when you use the `cancelOutput` argument?

Observers and Outputs

Observers and outputs are terminal nodes in the reactive graph. They differ from reactive expressions in two important ways:

- They are eager and forgetful—they run as soon as they possibly can, and they don't remember their previous action. This eagerness is "infectious" because if they use a reactive expression, that reactive expression will also be evaluated.

- The value returned by an observer is ignored because they are designed to work with functions called for their side effects, like `cat()` or `write.csv()`.

Observers and outputs are powered by the same underlying tool: `observe()`. This sets up a block of code that is run every time one of the reactive values or expressions it uses is updated. Note that the observer runs immediately when you create it—it must do this in order to determine its reactive dependencies:

```r
y <- reactiveVal(10)
observe({
  message("`y` is ", y())
})
#> Warning: Error in y: could not find function "y"

y(5)
y(4)
```

I rarely use `observe()` in this book, because it's the low-level tool that powers the user-friendly `observeEvent()`. Generally, you should stick with `observeEvent()` unless it's impossible to get it to do what you want. In this book, I'll only show you one case where `observe()` is necessary, "Pausing Animations" on page 239.

`observe()` also powers reactive outputs. Reactive outputs are a special type of observer that has two important properties:

- They are defined when you assign them into `output` (i.e., `output$text <- ...` creates the observer).
- They have some limited ability to detect when they're not visible (i.e., they're in nonactive tab) so they don't have to recompute.[3]

It's important to note that `observe()` and the reactive outputs don't "do" something but "create" something (which then takes action as needed). That mindset helps you to understand what's going on in this example:

```r
x <- reactiveVal(1)
y <- observe({
  x()
  observe(print(x()))
})
#> Warning: Error in x: could not find function "x"
x(2)
x(3)
```

---

3 In rare cases, you may prefer to process even outputs that are hidden. You can use the `outputOptions()` function's `suspendWhenHidden` to opt out of the automatic suspension feature on an output-by-output basis.

Each change to x causes the observer to be triggered. The observer itself calls observe(), setting up *another* observer. So each time x changes, it gets another observer, so its value is printed another time.

As a general rule, you should only ever create observers or outputs at the top level of your server function. If you find yourself trying to nest them or create an observer inside an output, sit down and sketch out the reactive graph that you're trying to create—there's almost certainly a better approach. It can be harder to spot this mistake directly in a more complex app, but you can always use the reactlog: just look for unexpected churn in observers (or outputs), then track back to what is creating them.

Isolating Code

To finish off the chapter, I will discuss two important tools for controlling exactly how and when the reactive graph is invalidated. In this section, I'll discuss isolate(), the tool that powers observeEvent() and eventReactive() and that lets you avoid creating reactive dependencies when not needed. In the next section, you'll learn about invalidateLater(), which allows you to generate reactive invalidations on a schedule.

isolate()

Observers are often coupled with reactive values in order to track state changes over time. For example, take this code that tracks how many times x changes:

```
r <- reactiveValues(count = 0, x = 1)
observe({
  r$x
  r$count <- r$count + 1
})
```

If you were to run it, you'd immediately get stuck in an infinite loop because the observer will take a reactive dependency on x *and* count; and since the observer modifies count, it will immediately rerun.

Fortunately, Shiny provides isolate() to resolve this problem. This function allows you to access the current value of a reactive value or expression *without* taking a dependency on it:

```
r <- reactiveValues(count = 0, x = 1)
class(r)
#> [1] "rv_flush_on_write" "reactivevalues"
observe({
  r$x
  r$count <- isolate(r$count) + 1
})
#> Warning: Error in <observer>: object 'r' not found
```

```
r$x <- 1
r$x <- 2
r$count
#> [1] 0

r$x <- 3
r$count
#> [1] 0
```

Like observe(), a lot of the time you don't need to use isolate() directly because there are two useful functions that wrap up the most common usage: observeE vent() and eventReactive().

observeEvent() and eventReactive()

When you saw the preceding code, you might have remembered "Observers" on page 49 and wondered why I didn't use observeEvent():

```
observeEvent(x(), {
  count(count() + 1)
})
```

And indeed, I could have because observeEvent(x, y) is equivalent to observe({x; isolate(y)}). It elegantly decouples what you want to listen to from what action you want to take. And eventReactive() performs the analogous job for reactives: eventReactive(x, y) is equivalent to reactive({x; isolate(y)}).

observeEvent() and eventReactive() have additional arguments that allow you to control the details of their operation:

- By default, both functions will ignore any event that yields NULL (or, in the special case of action buttons, 0). Use ignoreNULL = FALSE to also handle NULL values.
- By default, both functions will run once when you create them. Use ignoreInit = TRUE to skip this run.
- For observeEvent() only, you can use once = TRUE to run the handler only once.

These are rarely needed but good to know about so that you can look up the details from the documentation when you need them.

Exercises

1. Complete the following app with a server function that updates out with the value of x only when the button is pressed:

```
ui <- fluidPage(
  numericInput("x", "x", value = 50, min = 0, max = 100),
  actionButton("capture", "capture"),
  textOutput("out")
)
```

Timed Invalidation

`isolate()` reduces the times the reactive graph is invalidated. The topic of this section, `invalidateLater()`, does the opposite: it lets you invalidate the reactive graph when no data has changed. You saw an example of this in "Timed Invalidation" on page 45 with `reactiveTimer()`, but the time has come to discuss the underlying tool that powers it: `invalidateLater()`.

`invalidateLater(ms)` causes any reactive consumer to be invalidated in the future, after `ms` (milliseconds). It is useful for creating animations and connecting to data sources outside of Shiny's reactive framework that may be changing over time. For example, this reactive will automatically generate 10 fresh random normals every half a second:[4]

```
x <- reactive({
  invalidateLater(500)
  rnorm(10)
})
```

And this observer will increment a cumulative sum with a random number:

```
sum <- reactiveVal(0)
observe({
  invalidateLater(300)
  sum(isolate(sum()) + runif(1))
})
```

In the following sections, you'll learn how to use `invalidateLater()` to read changing data from disk, how to avoid getting `invalidateLater()` stuck in an infinite loop, and some occasionally important details of exactly when the invalidation happens.

---

4 Assuming that it's used by some output or observer; otherwise, it will stay in its initial invalidated state forever.

Polling

A useful application of `invalidateLater()` is to connect Shiny to data that is changing outside of R. For example, you could use the following reactive to reread a CSV file every second:

```
data <- reactive({
  on.exit(invalidateLater(1000))
  read.csv("data.csv")
})
```

This connects changing data into Shiny's reactive graph, but it has a serious downside: when you invalidate the reactive, you're also invalidating all downstream consumers, so even if the data is the same, all the downstream work has to be redone.

To avoid this problem, Shiny provides `reactivePoll()`, which takes two functions: one that performs a relatively cheap check to see if the data has changed, and another more expensive function that actually does the computation. We can use `reactivePoll()` to rewrite the previous reactive:

```
server <- function(input, output, session) {
  data <- reactivePoll(1000, session,
    function() file.mtime("data.csv"),
    function() read.csv("data.csv")
  )
}
```

Here we used `file.mtime()`, which returns the last time the file was modified, as a cheap check to see if we needed to reload the file.

Reading a file when it changes is a common task, so Shiny provides an even more specific helper that just needs a filename and a reader function:

```
server <- function(input, output, session) {
  data <- reactiveFileReader(1000, session, "data.csv", read.csv)
}
```

If you need to read changing data from other sources (e.g., a database), you'll need to come up with your own `reactivePoll()` code.

Long-Running Reactives

If you're performing a long-running computation, there's an important question you need to consider: When should you execute `invalidateLater()`? For example, take this reactive:

```
x <- reactive({
  invalidateLater(500)
  Sys.sleep(1)
  10
})
```

Assume Shiny starts the reactive running at time 0 and will request invalidation at time 500. The reactive takes 1000 ms to run, so it's now time 1000, and it's immediately invalidated and must be recomputed, which then sets up another invalidation: we're stuck in an infinite loop.

On the other hand, if you run `invalidateLater()` at the end, it will invalidate 500 ms after completion, so the reactive will be rerun every 1500 ms:

```
x <- reactive({
  on.exit(invalidateLater(500), add = TRUE)
  Sys.sleep(1)
  10
})
```

This is the main reason to prefer `invalidateLater()` to the simpler `reactive Timer()` that we used earlier: it gives you greater control over exactly when the invalidation occurs.

Timer Accuracy

The number of milliseconds specified in `invalidateLater()` is a polite request, not a demand. R may be doing other things when you asked for invalidation to occur, so your request has to wait. This effectively means that the number is a minimum and invalidation might take longer than you expect. In most cases, this doesn't matter because small differences are unlikely to affect user perception of your app. However, in situations where many small errors will accumulate, you should compute the exact elapsed time and use it to adjust your calculations.

For example, the following code computes distance based on velocity and elapsed time. Rather than assuming `invalidateLater(100)` always delays by exactly 100 ms, I compute the elapsed time and use it in my calculation of position:

```
velocity <- 3
r <- reactiveValues(distance = 1)

last <- proc.time()[[3]]
observe({
  cur <- proc.time()[[3]]
  time <- last - cur
  last <<- cur

  r$distance <- isolate(r$distance) + velocity * time
  invalidateLater(100)
})
```

If you're not doing careful animation, feel free to ignore the inherent variation in `invalidateLater()`. Just remember that it's a polite request, not a demand.

Exercises

1. Why will this reactive never be executed? Your explanation should talk about the reactive graph and invalidation:

```
server <- function(input, output, session) {
  x <- reactive({
    invalidateLater(500)
    rnorm(10)
  })
}
```

2. If you're familiar with SQL, use `reactivePoll()` to only reread an imaginary "Results" table whenever a new row is added. You can assume the Results table has a `timestamp` field that contains the date-time that a record was added.

Summary

In this chapter you've learned more about the building blocks that make Shiny work: reactive values, reactive expressions, observers, and timed evaluation. Now we'll turn our attention to a specific combination of reactive values and observers that allows us to escape some of the constraints (for better and worse) of the reactive graph.

## CHAPTER 16

Escaping the Graph

Introduction

Shiny's reactive programming framework is incredibly useful because it automatically determines the minimal set of computations needed to update all outputs when an input changes. But this framework is deliberately constraining, and sometimes you need to break free to do something risky but necessary.

In this chapter you'll learn how you can combine `reactiveValues()` and `observe()`/`observeEvent()` to connect the right-hand side of the reactive graph back to the left-hand side. These techniques are powerful because they give you manual control over parts of the graph. But they're also dangerous because they allow your app to do unnecessary work. Most importantly, you can now create infinite loops where your app gets stuck in a cycle of updates that never ends.

If you find the ideas explored in this chapter to be interesting, you might also want to look at the shinySignals (*https://oreil.ly/nvsID*) and rxtools (*https://oreil.ly/eCqn3*) packages. These are both experimental packages designed to explore "higher order" reactivity, reactives that are created programmatically from other reactives. I wouldn't recommend you use them in "real" apps, but reading the source code might be illuminating. To begin, we'll load shiny:

```
library(shiny)
```

What Doesn't the Reactive Graph Capture?

In "An Input Changes" on page 215, we discussed what happens when the user causes an input to be invalidated. There are two other important cases where you as the app author might invalidate an input:

- When you call an `update` function setting the `value` argument. This sends a message to the browser to change the value of an input, which then notifies R that the input value has been changed.

- When you modify the value of a reactive value (created with `reactiveVal()` or `reactiveValues()`).

It's important to understand that in both of these cases a reactive dependency is *not* created between the reactive value and the observer. While these actions cause the graph to invalidate, they are not recorded through new connections.[1]

To make this idea concrete, take the following simple app, with reactive graph shown in Figure 16-1:

```r
ui <- fluidPage(
  textInput("nm", "name"),
  actionButton("clr", "Clear"),
  textOutput("hi")
)
server <- function(input, output, session) {
  hi <- reactive(paste0("Hi ", input$nm))
  output$hi <- renderText(hi())
  observeEvent(input$clr, {
    updateTextInput(session, "nm", value = "")
  })
}
```

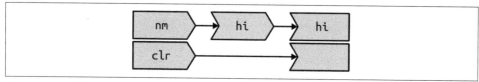

Figure 16-1. The reactive graph does not record the connection between the unnamed observer and the nm input; this dependency is outside of its scope.

What happens when you press the clear button?

1. `input$clr` invalidates, which then invalidates the observer.

2. The observer recomputes, re-creating the dependency on `input$clr` and telling the browser to change the value of the input control.

3. The browser changes the value of `nm`.

4. `input$nm` invalidates, invalidating `hi()` then `output$hi`.

---

1 As a debugging aid, the reactlog package can capture and draw these connections when you modify reactive values from an observer, but this information is not used by Shiny.

5. output$hi recomputes, forcing hi() to recompute.

None of these actions change the reactive graph, so it remains as in Figure 16-1, and the graph does not capture the connection from the observer to input$nm.

Case Studies

Next, let's take a look at a few cases where you might combine reactiveValues() and observeEvent() or observe() to solve problems that are otherwise very challenging (if not impossible). These are useful templates for your own apps.

One Output Modified by Multiple Inputs

To get started, we'll tackle a very simple problem: I want a common text box that's updated by multiple events:[2]

```
ui <- fluidPage(
  actionButton("drink", "drink me"),
  actionButton("eat", "eat me"),
  textOutput("notice")
)
server <- function(input, output, session) {
  r <- reactiveValues(notice = "")
  observeEvent(input$drink, {
    r$notice <- "You are no longer thirsty"
  })
  observeEvent(input$eat, {
    r$notice <- "You are no longer hungry"
  })
  output$notice <- renderText(r$notice)
}
```

Things get slightly more complicated in the next example, where we have an app with two buttons that let you increase and decrease values. We use a reactiveValues() to store the current value, and then use observeEvent() to increment and decrement the value when the appropriate button is pushed. The main additional complexity here is that the new value of n() depends on the previous value:

```
ui <- fluidPage(
  actionButton("up", "up"),
  actionButton("down", "down"),
  textOutput("n")
)
server <- function(input, output, session) {
  r <- reactiveValues(n = 0)
  observeEvent(input$up, {
```

---

2 This is rather similar to a notification, as seen in "Notifications" on page 126.

```
    r$n <- r$n + 1
  })
  observeEvent(input$down, {
    r$n <- r$n - 1
  })

  output$n <- renderText(r$n)
}
```

Figure 16-2 shows the reactive graph for this example. Again, note that the reactive graph does not include any connection from the observers back to the reactive value n.

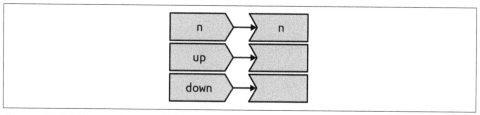

Figure 16-2. The reactive graph does not capture connections from observers to input values.

Accumulating Inputs

It's a similar pattern if you want to accumulate data in order to support data entry. Here the main difference is that we use updateTextInput() to reset the text box after the user clicks the add button:

```
ui <- fluidPage(
  textInput("name", "name"),
  actionButton("add", "add"),
  textOutput("names")
)
server <- function(input, output, session) {
  r <- reactiveValues(names = character())
  observeEvent(input$add, {
    r$names <- c(input$name, r$names)
    updateTextInput(session, "name", value = "")
  })

  output$names <- renderText(r$names)
}
```

We could make this slightly more useful by providing a delete button and making sure that the add button doesn't create duplicate names:

```
ui <- fluidPage(
  textInput("name", "name"),
  actionButton("add", "add"),
  actionButton("del", "delete"),
```

```
    textOutput("names")
)
server <- function(input, output, session) {
  r <- reactiveValues(names = character())
  observeEvent(input$add, {
    r$names <- union(r$names, input$name)
    updateTextInput(session, "name", value = "")
  })
  observeEvent(input$del, {
    r$names <- setdiff(r$names, input$name)
    updateTextInput(session, "name", value = "")
  })

  output$names <- renderText(r$names)
}
```

Pausing Animations

Another common use case is to provide a pair of start and stop buttons that lets you control some recurring event. This example uses a running reactive value to control whether or not the number increments, and invalidateLater() to ensure that the observer is invalidated every 250 ms when running:

```
ui <- fluidPage(
  actionButton("start", "start"),
  actionButton("stop", "stop"),
  textOutput("n")
)
server <- function(input, output, session) {
  r <- reactiveValues(running = FALSE, n = 0)

  observeEvent(input$start, {
    r$running <- TRUE
  })
  observeEvent(input$stop, {
    r$running <- FALSE
  })

  observe({
    if (r$running) {
      r$n <- isolate(r$n) + 1
      invalidateLater(250)
    }
  })
  output$n <- renderText(r$n)
}
```

Notice in this case we can't easily use observeEvent() because we perform different actions depending on whether running() is TRUE or FALSE. We must instead use iso late(). If we don't, this observer would also take a reactive dependency on n, which it updates, so it would get stuck in an infinite loop.

Hopefully these examples start to give you a flavor of what programming with these functions feels like. It's very imperative: when this happens, do that; when that happens, do the other thing. This makes it easier to understand on a small scale but harder to understand when bigger pieces start interacting. So generally, you'll want to use this as sparingly as possible and keep it isolated so that the smallest possible number of observers modify the reactive value.

Exercises

1. Provide a server function that draws a histogram of one hundred random numbers from a normal distribution when Normal is clicked and one hundred random uniforms when Uniform is clicked:

```
ui <- fluidPage(
  actionButton("rnorm", "Normal"),
  actionButton("runif", "Uniform"),
  plotOutput("plot")
)
```

2. Modify the preceding code to work with this UI:

```
ui <- fluidPage(
  selectInput("type", "type", c("Normal", "Uniform")),
  actionButton("go", "go"),
  plotOutput("plot")
)
```

3. Rewrite your code from the previous answer to eliminate the use of `observe()`/`observeEvent()` and only use `reactive()`. Why can you do that for the second UI but not the first?

Antipatterns

Once you get the hang of this pattern, it's easy to fall into bad habits:

```
server <- function(input, output, session) {
  r <- reactiveValues(df = cars)
  observe({
    r$df <- head(cars, input$nrows)
  })

  output$plot <- renderPlot(plot(r$df))
  output$table <- renderTable(r$df)
}
```

In this simple case, this code doesn't do much extra work compared to the alternative that uses `reactive()`:

```
server <- function(input, output, session) {
  df <- reactive(head(cars, input$nrows))

  output$plot <- renderPlot(plot(df()))
  output$table <- renderTable(df())
}
```

But there are still two drawbacks:

- If the table or plot are in tabs that are not currently visible, the observer will still draw/plot them.
- If the `head()` throws an error, the `observe()` will terminate the app, but the `reactive()` will propagate it so its displayed reactive throws an error and it won't get propagated.

And things will get progressively worse as the app gets more complicated. It's very easy to revert to the event-driven programming situation described in "Event-Driven Programming" on page 203. You end up doing a lot of hard work to analyze the flow of events in your app rather than relying on Shiny to handle it for you automatically.

It's informative to compare the two reactive graphs. Figure 16-3 shows the graph from the first example. It's misleading because it doesn't look like `nrows` is connected to `df()`. Using a reactive, as in Figure 16-4, makes the precise connection easy to see. Having a reactive graph that is as simple as possible is important for both humans and for Shiny. A simple graph is easier for humans to understand, and a simple graph is easier for Shiny to optimize.

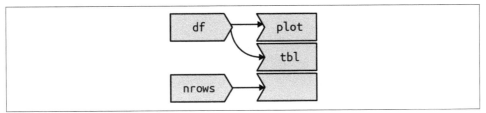

Figure 16-3. Using reactive values and observers leaves part of the graph disconnected.

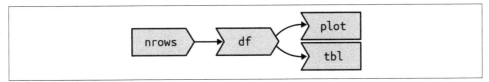

Figure 16-4. Using a reactive makes the dependencies between the components very clear.

Summary

In the last four chapters, you have learned much more about the reactive programming model used by Shiny. You've learned why reactive programming is important (it allows Shiny to do just as much work as is required and no more) and the details of the reactive graph. You've also learned a bit about how the fundamental building blocks work under the hood and how you can use them to escape the constraints of the reactive graph when needed.

The remainder of the book discusses Shiny through the lens of software engineering. In the next seven chapters, you'll learn how to keep your Shiny apps maintainable, performant, and safe as they continue to grow in size and impact.

## PART IV

Best Practices

When you start using Shiny, it'll take you a long time to make even small apps, because you have to learn the fundamentals. Over time, however, you'll become more comfortable with the basic interface of the package and the key ideas of reactivity, and you'll be able to create larger, more complex applications. As you start to write larger apps, you'll encounter a new set of challenges: keeping a complex and growing codebase organized, stable, and maintainable. This will include problems like:

- "I can't find the code I'm looking for in this huge file."
- "I haven't worked on this code in six months, and I'm afraid I'm going to break it if I make any changes."
- "Someone else started working with me on the application, and we keep standing on each other's toes."
- "The app works on my computer but doesn't work on my collaborator's or in production."

In this, the "best practices," part of the book, you'll learn some key concepts and tools from software engineering that will help you overcome these challenges:

- In Chapter 17, I'll briefly introduce you to the big ideas of software engineering.
- In Chapter 18, I'll show you how to extract code out of your Shiny app into independent apps and discuss why you might want to do so.
- In Chapter 19, you'll learn about Shiny's module system, which allows you to extract coupled UI and server code into isolated and reusable components.

- In Chapter 20, I'll show you how to turn your app in R package and show you why that investment will pay off for bigger apps.
- In Chapter 21, you'll learn how to turn your existing informal tests into automated tests that can easily be rerun whenever your app changes.
- In Chapter 22, you'll learn the patterns to avoid to keep your app secure.
- In Chapter 23, you'll learn how to identify and resolve performance bottlenecks in your apps, ensuring they remain speedy even when used by hundreds of users.

Of course, you can't learn everything about software engineering in one part of one book, so I'll also point you to good places to learn more.

## CHAPTER 17

General Guidelines

Introduction

This chapter introduces the most important software engineering skills you'll need when writing Shiny apps: code organization, testing, dependency management, source code control, continuous integration, and code reviews. These skills are not specific to Shiny apps, but you'll need to learn a bit about all of them if you want to write complex apps that get easier to maintain over time, not harder.

Improving your software engineering skills is a lifelong journey. Expect to have frustrations as you start learning them, but understand that everyone experiences the same issues, and if you persevere you'll get past them. Most people go through the same evolution when learning a new technique: "I don't understand it and have to look it up every time I use it" to "I vaguely understand it but still read the documentation a lot" to eventually "I understand it and can use it fluidly." It takes time and practice to get to the final stage.

I recommend setting aside some time each week to practice your software development skills. During this time, try to avoid touching the behavior or appearance of your app and instead focus your efforts on making the app easier to understand and develop. This will make your app easier to change in the future, and as you improve your software development skills, your first attempt at an app will also become higher quality.

(Thanks to my colleague Jeff Allen for contributing the bulk of this chapter.)

Code Organization

Any fool can write code that a computer can understand. Good programmers write code that humans can understand.

—Martin Fowler, *Refactoring: Improving the Design of Existing Code*

One of the most obvious ways to improve the quality of an application is to improve the readability and understandability of its code. The best programmers in the world can't maintain a codebase that they can't understand, so this is a good place to start.

Being a good programmer means developing empathy for others who will need to interact with this codebase in the future (even if it's just future-you!). Like all forms of empathy, this takes practice and becomes easier only after you've done it many times. Over time, you'll start to notice that certain practices improve the readability of your code. There are no universal rules, but some general guidelines include:

- Are the variable and function names clear and concise? If not, what names would better communicate the intent of the code?

- Do I have comments where needed to explain complex bits of code?

- Does this whole function fit on my screen, or could it be printed on a single piece of paper? If not, is there a way to break it up into smaller pieces?

- Am I copying and pasting the same block of code many times throughout my app? If so, is there a way to use a function or a variable to avoid the repetition?

- Are all the parts of my application tangled together, or can I manage the different components of my application in isolation?

There's no silver bullet to address all of these points—and many times they involve subjective judgment calls—but there are two particularly important tools:

Functions
: The topic of Chapter 18, functions allow you to reduce duplication in your UI code, make your server functions easier to understand and test, and allow you to more flexibly organize your app code.

Shiny modules
: The topic of Chapter 19, Shiny modules make it possible to write isolated, reusable code that coordinates frontend and backend behavior. Modules allow you to gracefully separate concerns so that (for example) individual pages in your application can operate independently or repeated components no longer need to be copied and pasted.

Testing

Developing a test plan for an application is critical to ensure its ongoing stability. Without a test plan, every change jeopardizes the application. When the application is small enough that you can hold it all in your head, you might feel that there's no need for an additional test plan. And sure, testing very simple apps can seem like more trouble than it's worth. However, the lack of a plan is likely to cause pain as soon as someone else starts contributing to your app, or when you've spent enough time away from it that you've forgotten how it all fits together.

A testing plan could be entirely manual. A great place to start is a simple text file giving a script to follow to check that all is well. However, that script will have to grow as the application becomes more complex, and you'll either spend more and more of your time manually testing the application or start skipping some of the script.

So the next step is to start to automate some of your testing. Automation takes time to set up, but it pays off over time because you can run the tests more frequently. For that reason, various forms of automated testing have been developed for Shiny, as outlined in Chapter 21. As that chapter will explain, you can develop:

- Unit tests that confirm the correct behavior of an individual function
- Integration tests to confirm the interactions between reactives
- Functional tests to validate the end-to-end experience from a browser
- Load tests to ensure that the application can withstand the amount of traffic you anticipate for it

The beauty of writing an automated test is that once you've taken the time to write it, you'll never need to manually test that portion of the application again. You can even leverage continuous integration (more on that shortly) to run these tests every time you make a change to your code before publishing the application.

Dependency Management

If you've ever tried to reproduce some analysis in R written by someone else, or even tried to rerun some analysis or Shiny application you wrote some time ago, you may have run into trouble around dependencies. An app's dependencies are anything beyond the source code that it requires to run. These could include files on the hard drive, an external database or API, or other R packages that are used by the app.

For any analysis that you may want to reproduce in the future, consider using renv (*https://rstudio.github.io/renv*), which enables you to create **r**eproducible R **e**nvironments. Using renv, you can capture the exact package versions that your application uses so that when you go to use this application on another computer, you can use exactly the same package versions. This is vital for apps run in production, not just because it gets the versions right on the first run but because it also isolates your app from version changes over time.

Another tool for managing dependencies is the config package (*https://github.com/rstudio/config*). The config package doesn't actually manage dependencies itself, but it does provide a convenient place for you to track and manage dependencies other than R packages. For instance, you might specify the path to a CSV file that your application depends on or the URL of an API that you require. Having these enumerated in the config file gives you a single place where you can track and manage these dependencies. Even better, it enables you to create different configurations for different environments. For example, if your application analyzes a database with lots of data, you might choose to configure a few different environments:

- In the production environment, you connect the app to the real "production" database.

- In a test environment, you can configure the app to use a test database so that you properly exercise the database connections in your tests, but you don't risk corrupting your production database if you accidentally make a change that corrupts the data.

- In development, you might configure the application to use a small CSV with a subset of data to allow for faster iterating.

Lastly, be wary of making assumptions about the local filesystem. If your code has references to data at *C:\data\cars.csv* or *~/my-projects/genes.rds*, for example, you need to realize that it's very unlikely that these files will exist on another computer. Instead, either use a path relative to the app directory (e.g., *data/cars.csv* or *genes.rds*), or use the config package to make the external path explicit and configurable.

Source Code Management

Anyone who's been programming for a long time has inevitably arrived at a state where they've accidentally broken their app and want to roll back to a previous working state. This is incredibly arduous when done manually. Fortunately, however, you can rely on a "version-control system" that makes it easy to track atomic changes, roll back to previous work, and integrate the work of multiple contributors.

The most popular version-control system in the R community is Git. Git is typically paired with GitHub, a website that makes it easy to share your Git repos with others. It definitely takes work to become proficient with Git and GitHub, but any experienced developer will confirm that the effort is well worth it. If you're new to Git, I'd highly recommend starting with *Happy Git and GitHub for the useR* (*https://happygit withr.com*), by Jenny Bryan.

Continuous Integration/Deployment

Once you are using a version control system and have a robust set of automated tests, you might benefit from continuous integration (CI). CI is a way to perpetually validate that the changes you're making to your application haven't broken anything. You can use it retroactively (to notify you if a change you just made broke your application) or proactively (to notify you if a *proposed* change would break your app).

There are a variety of services that can connect to a Git repo and automatically run tests when you push a new commit or propose changes. Depending on where your code is hosted, you can consider GitHub actions (*https://github.com/features/actions*), Travis CI (*https://travis-ci.org*), Azure Pipelines (*https://oreil.ly/NFbFQ*), AppVeyor (*https://www.appveyor.com*), Jenkins (*https://jenkins.io*), or GitLab CI/CD (*https://oreil.ly/WYsfG*), to name a few.

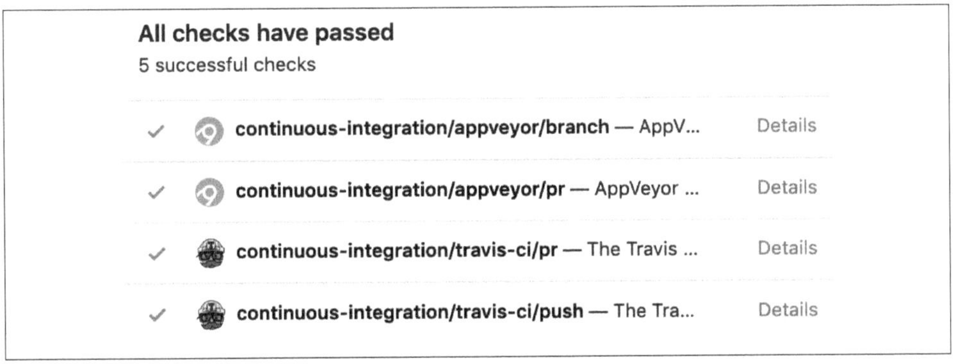

Figure 17-1. An example CI run, showing successful results across four independent testing environments.

Figure 17-1 shows what this looks like when a CI system is connected to GitHub to test pull requests. As you can see, all the CI tests show green checks, meaning that each of the automated test environments were successful. If any of the tests had failed, you would be alerted to the failure before you merge the changes into your app. Having a CI process not only prevents experienced developers from making accidental mistakes but also helps new contributors feel confident in their changes.

Code Reviews

Many software companies have found the benefits of having someone else review code before it's formally incorporated into a codebase. This process of "code review" has a number of benefits:

- It catches bugs before they get incorporated into the application, making them much less expensive to fix.

- It offers teaching opportunities—programmers at all levels often learn something new by reviewing others' code or by having their code reviewed.

- It facilitates cross-pollination and knowledge sharing across a team to eliminate having only one person who understands the app.

- The resulting conversation often improves the readability of the code.

Typically, a code review involves someone other than you, but you can still benefit even if it's only you. Most experienced developers will agree that taking a moment to review your own code often reveals some small flaw, particularly if you can let it sit for at least a few hours between writing and review.

Here are few questions to hold in your head when reviewing code:

- Do new functions have concise but evocative names?

- Are there parts of the code you find confusing?

- What areas are likely to change in the future and would particularly benefit from automated testing?

- Does the style of the code match the rest of the app (or, even better, your group's documented code style)?

If you're embedded in an organization with a strong engineering culture, setting up code reviews for data science code should be relatively straightforward, and you'll have existing tools and experience to draw on. If you're in an organization that has few other software engineers, you may need to do more convincing.

Two resources I'd recommend:

- thoughtbot guides (*https://oreil.ly/949mo*)

- Code Review Developer Guide (*https://oreil.ly/VZEpM*)

Summary

Now that you've learned a little bit of the software engineer mindset, the next chapters are going to dive into the details of function writing, testing, security, and performance as they apply to Shiny apps. You'll need to read Chapter 18 before the other chapters, but otherwise you can skip around.

Functions

As your app gets bigger, it will get harder and harder to hold all the pieces in your head, making it harder and harder to understand. In turn, this makes it harder to add new features and harder to find a solution when something goes wrong (i.e., it's harder to debug). If you don't take deliberate steps, the development pace of your app will slow, and it will become less and less enjoyable to work on.

In this chapter, you'll learn how writing functions can help. This tends to have slightly different flavors for UI and server components:

- In the UI, you have components that are repeated in multiple places with minor variations. Pulling out repeated code into a function reduces duplication (making it easier to update many controls from one place) and can be combined with functional programming techniques to generate many controls at once.

- In the server, complex reactives are hard to debug because you need to be in the midst of the app. Pulling out a reactive into a separate function, even if that function is only called in one place, makes it substantially easier to debug, because you can experiment with computation independent of reactivity.

Functions have another important role in Shiny apps: they allow you to spread out your app code across multiple files. While you certainly can have one giant *app.R* file, it's much easier to manage when spread across multiple files.

I assume that you're already familiar with the basics of functions.[1] The goal of this chapter is to activate your existing skills, showing you some specific cases where using functions can substantially improve the clarity of your app. Once you've

---

[1] If you're not, and you'd like to learn the basics, you might try reading the Functions chapter (*https://oreil.ly/yrvYN*) of *R for Data Science*.

mastered the ideas in this chapter, the next step is to learn how to write code that requires coordination across the UI and server. That requires *modules*, which you'll learn about in Chapter 19. As usual, we begin by loading shiny:

```
library(shiny)
```

File Organization

Before we go on to talk about exactly how you might use functions in your app, I want to start with one immediate benefit: functions can live outside of *app.R*. There are two places you might put them, depending on how big they are:

- I recommend putting large functions (and any smaller helper functions that they need) into their own *R/{function-name}.R* file.
- You might want to collect smaller, simpler, functions into one place. I often use *R/utils.R* for this, but if they're primarily used in your UI, you might use *R/ui.R*.

If you've made an R package before, you might notice that Shiny uses the same convention for storing files containing functions. And indeed, if you're making a complicated app, particularly if there are multiple authors, there are substantial advantages to making a full-fledged package. If you want to do this, I recommend reading *Engineering Production-Grade Shiny Apps* (*https://engineering-shiny.org*) and using the accompanying golem (*https://thinkr-open.github.io/golem*) package. We'll touch on packages again when we talk more about testing.

UI Functions

Functions are a powerful tool to reduce duplication in your UI code. Let's start with a concrete example of some duplicated code. Imagine that you're creating a bunch of sliders that each need to range from 0 to 1, starting at 0.5, with a 0.1 step. You *could* do a bunch of copy and paste to generate all the sliders:

```
ui <- fluidRow(
  sliderInput("alpha", "alpha", min = 0, max = 1, value = 0.5, step = 0.1),
  sliderInput("beta",  "beta",  min = 0, max = 1, value = 0.5, step = 0.1),
  sliderInput("gamma", "gamma", min = 0, max = 1, value = 0.5, step = 0.1),
  sliderInput("delta", "delta", min = 0, max = 1, value = 0.5, step = 0.1)
)
```

But I think it's worthwhile to recognize the repeated pattern and extract out a function. That makes the UI code substantially simpler:

```
sliderInput01 <- function(id) {
  sliderInput(id, label = id, min = 0, max = 1, value = 0.5, step = 0.1)
}

ui <- fluidRow(
```

```
sliderInput01("alpha"),
sliderInput01("beta"),
sliderInput01("gamma"),
sliderInput01("delta")
)
```

Here, a function helps in two ways:

- We can give the function an evocative name, making it easier to understand what's going on when we reread the code in the future.
- If we need to change the behavior, we only need to do it in one place. For example, if we decided that we needed a finer resolution for the steps, we only need to write `step = 0.01` in one place, not four.

Other Applications

Functions can be useful in many other places. Here are a few ideas to get your creative juices flowing:

- If you're using a customized `dateInput()` for your country, pull it out into one place so that you can use consistent arguments. For example, imagine you wanted a date control for Americans to use to select weekdays:

  ```
  usWeekDateInput <- function(inputId, ...) {
    dateInput(inputId, ..., format = "dd M, yy", daysofweekdisabled = c(0, 6))
  }
  ```

  Note the use of ...; it means that you can still pass along any other arguments to `dateInput()`.

- Or maybe you want a radio button that makes it easier to provide icons:

  ```
  iconRadioButtons <- function(inputId, label, choices, selected = NULL) {
    names <- lapply(choices, icon)
    values <- if (is.null(names(choices))) names(choices) else choices
    radioButtons(inputId,
      label = label,
      choiceNames = names, choiceValues = values, selected = selected
    )
  }
  ```

- Or if there are multiple selections you reuse in multiple places:

  ```
  stateSelectInput <- function(inputId, ...) {
    selectInput(inputId, ..., choices = state.name)
  }
  ```

If you're developing a lot of Shiny apps within your organization, you can help improve cross-app consistency by putting functions like this in a shared package.

Functional Programming

To return back to our motivating example, you could reduce the code still further if you're comfortable with functional programming:

```
library(purrr)

vars <- c("alpha", "beta", "gamma", "delta")
sliders <- map(vars, sliderInput01)
ui <- fluidRow(sliders)
```

There are two big ideas here:

- `map()` calls `sliderInput01()` once for each string stored in `vars`. It returns a list of sliders.
- When you pass a list into `fluidRow()` (or any HTML container), it automatically unpacks the list so that the elements become the children of the container.

If you would like to learn more about `map()` (or its base equivalent, `lapply()`), you might enjoy the Functionals chapter (*https://oreil.ly/Jd73D*) of *Advanced R*.

UI as Data

It's possible to generalize this idea further if the controls have more than one varying input. First, we create an inline data frame that defines the parameters of each control using `tibble::tribble()`. We're turning UI structure into an explicit data structure:

```
vars <- tibble::tribble(
  ~ id,      ~ min, ~ max,
  "alpha",      0,     1,
  "beta",       0,    10,
  "gamma",     -1,     1,
  "delta",      0,     1,
)
```

Then we create a function where the argument names match the column names:

```
mySliderInput <- function(id, label = id, min = 0, max = 1) {
  sliderInput(id, label, min = min, max = max, value = 0.5, step = 0.1)
}
```

Then finally we use `purrr::pmap()` to call `mySliderInput()` once for each row of `vars`:

```
sliders <- pmap(vars, mySliderInput)
```

Don't worry if this code looks like gibberish to you: you can continue to use copy and paste. But in the long run, I'd recommend learning more about functional programming, because it gives you such a wonderful ability to concisely express otherwise long-winded concepts. See "Creating UI with Code" on page 166 for more examples of using these techniques to generate a dynamic UI in response to user actions.

Server Functions

Whenever you have a long reactive (say > 10 lines), you should consider pulling it out into a separate function that does not use any reactivity. This has two advantages:

- It is much easier to debug and test your code if you can partition it so that reactivity lives inside of server() and complex computation lives in your functions.
- When looking at a reactive expression or output, there's no way to easily tell exactly what values it depends on, except by carefully reading the code block. A function definition, however, tells you exactly what the inputs are.

The key benefits of a function in the UI tend to be around reducing duplication. The key benefits of functions in a server tend to be around isolation and testing.

Reading Uploaded Data

Take this server from "Uploading Data" on page 143. It contains a moderately complex reactive():

```r
server <- function(input, output, session) {
  data <- reactive({
    req(input$file)

    ext <- tools::file_ext(input$file$name)
    switch(ext,
      csv = vroom::vroom(input$file$datapath, delim = ","),
      tsv = vroom::vroom(input$file$datapath, delim = "\t"),
      validate("Invalid file; Please upload a .csv or .tsv file")
    )
  })

  output$head <- renderTable({
    head(data(), input$n)
  })
}
```

If this was a real app, I'd seriously consider extracting out a function specifically for reading uploaded files:

```r
load_file <- function(name, path) {
  ext <- tools::file_ext(name)
  switch(ext,
    csv = vroom::vroom(path, delim = ","),
    tsv = vroom::vroom(path, delim = "\t"),
    validate("Invalid file; Please upload a .csv or .tsv file")
```

```
    )
  }
```

When extracting out such helpers, avoid taking reactives as input or returning outputs. Instead, pass values into arguments and assume the caller will turn the result into a reactive if needed. This isn't a hard-and-fast rule; sometimes it will make sense for your functions to input or output reactives. But generally, I think it's better to keep the reactive and nonreactive parts of your app as separate as possible. In this case, I'm still using `validate()`; that works because outside of Shiny, `validate()` works similarly to `stop()`. But I keep the `req()` in the server, because it shouldn't be the responsibility of the file parsing code to know when it's run.

Since this is now an independent function, it could live in its own file (*R/load_file.R*, say), keeping the `server()` svelte. This helps keep the server function focused on the big picture of reactivity rather than on the smaller details underlying each component:

```
server <- function(input, output, session) {
  data <- reactive({
    req(input$file)
    load_file(input$file$name, input$file$datapath)
  })

  output$head <- renderTable({
    head(data(), input$n)
  })
}
```

The other big advantage is that you can play with `load_file()` at the console, outside of your Shiny app. If you move toward formal testing of your app (see Chapter 21), this also makes that code easier to test.

Internal Functions

Most of the time you'll want to make the function completely independent of the server function so that you can put it in a separate file. However, if the function needs to use `input`, `output`, or `session`, it may make sense for the function to live inside the server function:

```
server <- function(input, output, session) {
  switch_page <- function(i) {
    updateTabsetPanel(input = "wizard", selected = paste0("page_", i))
  }

  observeEvent(input$page_12, switch_page(2))
  observeEvent(input$page_21, switch_page(1))
  observeEvent(input$page_23, switch_page(3))
  observeEvent(input$page_32, switch_page(2))
}
```

This doesn't make testing or debugging any easier, but it does reduce duplicated code.

We could, of course, add `session` to the arguments of the function:

```
switch_page <- function(i) {
  updateTabsetPanel(input = "wizard", selected = paste0("page_", i))
}

server <- function(input, output, session) {
  observeEvent(input$page_12, switch_page(2))
  observeEvent(input$page_21, switch_page(1))
  observeEvent(input$page_23, switch_page(3))
  observeEvent(input$page_32, switch_page(2))
}
```

But this feels weird as the function is still fundamentally coupled to this app because it only affects a control named "wizard" with a very specific set of tabs.

Summary

As your apps get bigger, extracting nonreactive functions out of the flow of the app will make your life substantially easier. Functions allow you to separate reactive and nonreactive code and spread your code out over multiple files. This often makes it much easier to see the big-picture shape of your app, and by moving complex logic out of the app into regular R code, it makes it much easier to experiment, iterate, and test. When you start extracting out function, it's likely to feel a bit slow and frustrating, but over time you'll get faster and faster, and soon it will become a key tool in your toolbox.

The functions in this chapter have one important drawback: they can generate only UI or server components, not both. In the next chapter, you'll learn how to create Shiny modules, which coordinate UI and server code into a single object.

## CHAPTER 19

Shiny Modules

In the last chapter we used functions to decompose parts of your Shiny app into independent pieces. Functions work well for code that is either completely on the server side or completely on the client side. For code that spans both (i.e., whether the server code relies on specific structure in the UI), you'll need a new technique: modules.

At the simplest level, a module is a pair of UI and server functions. The magic of modules comes because these functions are constructed in a special way that creates a "namespace." So far, when writing an app, the names (ids) of the controls are global: all parts of your server function can see all parts of your UI. Modules give you the ability to create controls that can only be seen from within the module. This is called a *namespace* because it creates "spaces" of "names" that are isolated from the rest of the app.

Shiny modules have two big advantages. First, namespacing makes it easier to understand how your app works because you can write, analyze, and test individual components in isolation. Second, because modules are functions, they help you reuse code; anything you can do with a function, you can do with a module. Let's begin by loading shiny:

```
library(shiny)
```

Motivation

Before we dive into the details of creating modules, it's useful to get a sense for how they change the "shape" of your app. I'm going to borrow an example from Eric Nantz (*https://github.com/rpodcast*), who talked about modules at rstudio::conf(2019) (*https://youtu.be/ylLLVo2VL50*). Eric was motivated to use modules because he had a big, complex app, as shown in Figure 19-1. You don't know the specifics of this app, but you can get some sense of the complexity due to the many interconnected components.

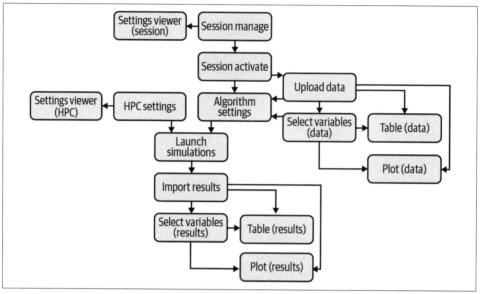

Figure 19-1. A rough sketch of a complex app. I've done my best to display it simply in a diagram, but it's still hard to understand what all the pieces are.

Figure 19-2 shows the how the app looks now, after a rewrite that uses modules:

- The app is divided up into pieces, and each piece has a name. Naming the pieces means that the names of the controls can be simpler. For example, previously the app had "session manage" and "session activate," but now we only need "manage" and "activate" because those controls are nested inside the session module. This is namespacing!

- A module is a black box with defined inputs and outputs. Other modules can only communicate via the interface (outside) of a module; they can't reach inside and directly inspect or modify the internal controls and reactives. This enforces a simpler structure to the whole app.

- Modules are reusable so we can write functions to generate both yellow and blue components. This can significantly reduce the total amount of code in the app.

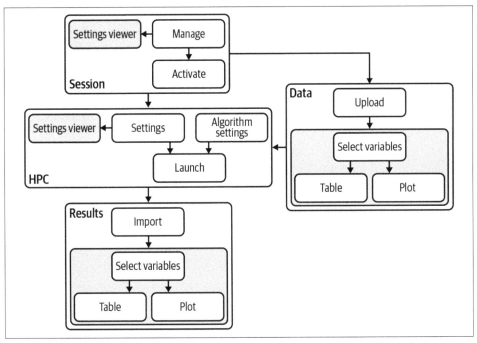

Figure 19-2. After converting the app to use modules, it's much easier to see the big-picture components of the app and see what is reused in multiple places (the blue and yellow components).

Module Basics

To create your first module, we'll pull a module out of a very simple app that draws a histogram:

```r
ui <- fluidPage(
  selectInput("var", "Variable", names(mtcars)),
  numericInput("bins", "bins", 10, min = 1),
  plotOutput("hist")
)
server <- function(input, output, session) {
  data <- reactive(mtcars[[input$var]])
  output$hist <- renderPlot({
    hist(data(), breaks = input$bins, main = input$var)
  }, res = 96)
}
```

This app is so simple that there's no real benefit to pulling out a module, but it will serve to illustrate the basic mechanics before we dive into more realistic, and hence complicated, use cases.

A module is very similar to an app. Like an app, it's composed of two pieces:[1]

- The *module UI* function that generates the ui specification
- The *module server* function that runs code inside the server function

The two functions have standard forms. They both take an id argument and use it to namespace the module. To create a module, we need to extract code out of the app UI and server and put it into the module UI and server.

Module UI

We'll start with the module UI. There are two steps:

- Put the UI code inside a function that has an id argument.
- Wrap each existing ID in a call to NS() so that (for example) "var" turns into NS(id, "var").

This yields the following function:

```
histogramUI <- function(id) {
  tagList(
    selectInput(NS(id, "var"), "Variable", choices = names(mtcars)),
    numericInput(NS(id, "bins"), "bins", value = 10, min = 1),
    plotOutput(NS(id, "hist"))
  )
}
```

Here I've returned the UI components in a tagList(), which is a special type of layout function that allows you to bundle together multiple components without actually implying how they'll be laid out. It's the responsibility of the person calling histogramUI() to wrap the result in a layout function like column() or fluidRow() according to their needs.

Module Server

Next we tackle the server function. This gets wrapped inside *another* function, which must have an id argument. This function calls moduleServer() with the id and a function that looks like a regular server function:

```
histogramServer <- function(id) {
  moduleServer(id, function(input, output, session) {
    data <- reactive(mtcars[[input$var]])
    output$hist <- renderPlot({
      hist(data(), breaks = input$bins, main = input$var)
```

---

1 Unlike an app, both module UI and server are functions.

```
  }, res = 96)
  })
}
```

The two levels of functions are important here. We'll come back to them later, but in short they help distinguish the argument to your module from the arguments to the server function. Don't worry if this looks very complex; it's basically boilerplate that you can copy and paste for each new module that you create.

Note that `moduleServer()` takes care of the namespacing automatically: inside of `moduleServer(id)`, `input$var` and `input$bins` refer to the inputs with names `NS(id, "var")` and `NS(id, "bins")`.

Updated App

Now that we have the UI and server functions, it's good practice to write a function that uses them to generate an app we can use for experimentation and testing:

```
histogramApp <- function() {
  ui <- fluidPage(
    histogramUI("hist1")
  )
  server <- function(input, output, session) {
    histogramServer("hist1")
  }
  shinyApp(ui, server)
}
```

Note that, like all Shiny control, you need to use the same `id` in both UI and server; otherwise the two pieces will not be connected.

Some Module History

Modules were introduced in Shiny 0.13 (January 2016) with `callModule()` and overhauled in Shiny 1.5.0 (June 2020) with the introduction of `moduleServer()`. If you've learned modules a while ago, you might have learned `callModule()` and be wondering what the deal is with `moduleServer()`. These two functions are identical, except that the first two arguments are flipped. This is a simple change that leads to a rather different structure for the entire app:

```
histogramServerOld <- function(input, output, session) {
  data <- reactive(mtcars[[input$var]])
  output$hist <- renderPlot({
    hist(data(), breaks = input$bins, main = input$var)
  }, res = 96)
}
server <- function(input, output, session) {
  callModule(histogramServerOld, "hist1")
}
```

The difference is largely superficial for this simple app, but moduleServer() makes more complicated modules with arguments considerably easier to understand.

Namespacing

Now that we have a complete app, let's circle back and talk about namespacing some more. The key idea that makes modules work is that the name of each control (i.e., its id) is now determined by two pieces:

- The first piece comes from the module *user*, the developer who calls histogram Server().

- The second piece comes from the module *author*, the developer who wrote histo gramServer().

This two-part specification means that you, the module author, don't need to worry about clashing with other UI components created by the user. You have your own "space" of names that you own and can arrange to best meet your own needs.

Namespacing turns modules into black boxes. From outside of the module, you can't see any of the inputs, outputs, or reactives inside of it. For example, take the following app. The text output output$out will never get updated because there is no input $bins; the bins input can only be seen inside of the hist1 module:

```
ui <- fluidPage(
  histogramUI("hist1"),
  textOutput("out")
)
server <- function(input, output, session) {
  histogramServer("hist1")
  output$out <- renderText(paste0("Bins: ", input$bins))
}
```

If you want to take input from reactives elsewhere in the app, you'll need to pass them to the module function explicitly; we'll come back to that shortly.

Note that the module UI and server differ in how the namespacing is expressed:

- In the module UI, the namespacing is *explicit*: you have to call NS(id, "name") every time you create an input or output.

- In the module server, the namespacing is *implicit*. You only need to use id in the call to moduleServer(). Shiny automatically namespaces input and output so that, in your module code, input$name means the input with name NS(id, "name").

Naming Conventions

In this example, I've used a special naming scheme for all the components of the module, and I recommend that you also use it for your own modules. Here, the module draws a histogram, so I've called it the histogram module. This base name is then used in a variety of places:

- *R/histogram.R* holds all the code for the module.
- histogramUI() is the module UI. If it's used primarily for input or output, I'd call it histogramInput() or histogramOutput() instead.
- histogramServer() is the module server.
- histogramApp() creates a complete app for interactive experimentation and more formal testing.

Exercises

1. Why is it good practice to put a module in its own file in the *R/* directory? What do you need to do to make sure it's loaded by your Shiny app?

2. The following module UI includes a critical mistake. What is it, and why will it cause problems?

   ```
   histogramUI <- function(id) {
     tagList(
       selectInput("var", "Variable", choices = names(mtcars)),
       numericInput("bins", "bins", value = 10, min = 1),
       plotOutput("hist")
     )
   }
   ```

3. The following module generates a new random number every time you click go:

   ```
   randomUI <- function(id) {
     tagList(
       textOutput(NS(id, "val")),
       actionButton(NS(id, "go"), "Go!")
     )
   }
   randomServer <- function(id) {
     moduleServer(id, function(input, output, session) {
       rand <- eventReactive(input$go, sample(100, 1))
       output$val <- renderText(rand())
     })
   }
   ```

   Create an app that displays four copies of this module on a single page. Verify that each module is independent. How could you change the return value of ran domUI() to make the display more attractive?

4. Are you sick of typing module boilerplate already? Read about RStudio snippets (*https://oreil.ly/yhvm1*) and add the following snippet to your RStudio config to make it even easier to create new modules:

```
${1}UI <- function(id) {
  tagList(
    ${2}
  )
}

${1}Server <- function(id) {
  moduleServer(id, function(input, output, session) {
    ${3}
  }
}
```

Inputs and Outputs

Sometimes a module with only an `id` argument to the module UI and server is useful because it allows you to isolate complex code in its own file. This is particularly useful for apps that aggregate independent components, such as a corporate dashboard where each tab shows tailored reports for each line of business. Here modules allow you to develop each piece in its own file without having to worry about IDs clashing across components.

A lot of the time, however, your module UI and server will need additional arguments. Adding arguments to the module UI gives greater control over module appearance, allowing you to use the same module in more places in your app. But the module UI is just a regular R function, so there's relatively little to learn that's specific to Shiny, and much of it was already covered in Chapter 18.

So in the following sections, I'll focus on the module server and discuss how your module can take additional reactive inputs and return one or more reactive outputs. Unlike regular Shiny code, connecting modules together requires you to be explicit about inputs and outputs. Initially, this is going to feel tiresome. And it's certainly more work than Shiny's usual free-form association. But modules enforce specific lines of communication for a reason: they're a little more work to create but much easier to understand and allow you to build substantially more complex apps.

You might see advice to use `session$userData` or other techniques to break out of the module straitjacket. Be wary of such advice: it's showing you how to work around the rules imposed by namespacing, making it easy to reintroduce much complexity to your app and significantly reducing the benefits of using a module in the first place.

Getting Started: UI Input and Server Output

To see how inputs and outputs work, we'll start off easy with a module that allows the user to select a dataset from built-in data provided by the datasets package. This isn't terribly useful by itself, but it illustrates some of the basic principles and is a useful building block for more complex modules, as you've seen before in "Adding UI Controls" on page 6.

We'll start with the module UI. Here I use a single additional argument so that you can limit the options to built-in datasets that are either data frames (filter = is.data.frame) or matrices (filter = is.matrix). I use this argument to optionally filter the objects found in the datasets package, then create a selectInput():

```
datasetInput <- function(id, filter = NULL) {
  names <- ls("package:datasets")
  if (!is.null(filter)) {
    data <- lapply(names, get, "package:datasets")
    names <- names[vapply(data, filter, logical(1))]
  }

  selectInput(NS(id, "dataset"), "Pick a dataset", choices = names)
}
```

The module server is also simple: we just use get() to retrieve the dataset with its name. There's one new idea here: like a function and unlike a regular server(), this module server returns a value. Here we take advantage of the usual rule that the last expression processed in the function becomes the return value.[2] This value should always be a reactive:

```
datasetServer <- function(id) {
  moduleServer(id, function(input, output, session) {
    reactive(get(input$dataset, "package:datasets"))
  })
}
```

To use a module server that returns something, you just have to capture its return value with <-. That's demonstrated in the following module app, where I capture the dataset and then display it in a tableOutput():

```
datasetApp <- function(filter = NULL) {
  ui <- fluidPage(
    datasetInput("dataset", filter = filter),
    tableOutput("data")
  )
  server <- function(input, output, session) {
    data <- datasetServer("dataset")
```

---

2 The tidyverse style guide (*https://oreil.ly/HFWiL*) recommends reserving return() only for cases where you are returning early.

```
    output$data <- renderTable(head(data()))
  }
  shinyApp(ui, server)
}
```

I've made a few executive decisions in my design of this function:

- It takes a `filter` argument that's passed along to the module UI, making it easy to experiment with that input argument.

- I use a tabular output to show all the data. It doesn't really matter what you use here, but the more expressive your UI, the easier it is to check that the module does what you expect.

Case Study: Selecting a Numeric Variable

Next, we'll create a control that allows the user to select variables of a specified type from a given reactive dataset. Because we want the dataset to be reactive, we can't fill in the choices when we start the app. This makes the module UI very simple:

```
selectVarInput <- function(id) {
  selectInput(NS(id, "var"), "Variable", choices = NULL)
}
```

The server function will have two arguments:

- The `data` to select variables from. I want this to be reactive so it can work with the `dataset` module I created previously.

- A `filter` used to select which variables to list. This will be set by the caller of the module, so it doesn't need to be reactive. To keep the module server simple, I've extracted out the key idea into a helper function:

```
find_vars <- function(data, filter) {
  names(data)[vapply(data, filter, logical(1))]
}
```

Then the module server uses `observeEvent()` to update the `inputSelect` choices when the data changes and returns a reactive that provides the values of the selected variable:

```
selectVarServer <- function(id, data, filter = is.numeric) {
  moduleServer(id, function(input, output, session) {
    observeEvent(data(), {
      updateSelectInput(session, "var", choices = find_vars(data(), filter))
    })

    reactive(data()[[input$var]])
  })
}
```

To make our app, we again capture the results of the module server and connect it to an output in our UI. I want to make sure all the reactive plumbing is correct, so I use the `dataset` module as a source of reactive data frames:

```
selectVarApp <- function(filter = is.numeric) {
  ui <- fluidPage(
    datasetInput("data", is.data.frame),
    selectVarInput("var"),
    verbatimTextOutput("out")
  )
  server <- function(input, output, session) {
    data <- datasetServer("data")
    var <- selectVarServer("var", data, filter = filter)
    output$out <- renderPrint(var())
  }

  shinyApp(ui, server)
}
```

Server Inputs

When designing a module server, you need to think about who is going to provide the value for each argument: is it the R programmer calling your module or the person using the app? Another way to think about this is when can the value change: is it fixed and constant over the lifetime of the app, or is it reactive, changing as the user interacts with the app? This is an important design decision that determines whether or not an argument should be a reactive or not.

Once you've made this decision, I think it's good practice to check that each input to your module is either reactive or constant. If you don't, and the user supplies the wrong type, they'll get a cryptic error message. You can make the life of a module user much easier with a quick and dirty call to `stopifnot()`. For example, `selectVarServer()` could check that `data` is reactive and `filter` is not with the following code:

```
selectVarServer <- function(id, data, filter = is.numeric) {
  stopifnot(is.reactive(data))
  stopifnot(!is.reactive(filter))

  moduleServer(id, function(input, output, session) {
    observeEvent(data(), {
      updateSelectInput(session, "var", choices = find_vars(data(), filter))
    })

    reactive(data()[[input$var]])
  })
}
```

If you expect the module to be used many times by many people, you might also consider hand-crafting an error message with an `if` statement and a call to `stop()`.

Checking that the module inputs are reactive (or not) helps you avoid a common problem when you mix modules with other input controls. input$var is not a reactive, so whenever you pass an input value into a module, you'll need to wrap it in a reactive() (e.g., selectVarServer("var", reactive(input$x))). If you check the inputs like I recommend here, you'll get a clear error; if you don't, you'll get something cryptic like could not find function "data".

 You might also apply this strategy to find_vars(). It's not quite as important here, but because debugging Shiny apps is a little harder than debugging regular R code, I think it does make sense to invest a little more time in checking inputs so that you get clearer error messages when something goes wrong:

```
find_vars <- function(data, filter) {
  stopifnot(is.data.frame(data))
  stopifnot(is.function(filter))
  names(data)[vapply(data, filter, logical(1))]
}
```

This caught a couple of errors that I made while working on this chapter.

Modules Inside of Modules

Before we continue on to talk more about outputs from your server function, I wanted to highlight that modules are composable, and it may make sense to create a module that itself contains a module. For example, we could combine the dataset and selectVar modules to make a module that allows the user to pick a variable from a built-in dataset:

```
selectDataVarUI <- function(id) {
  tagList(
    datasetInput(NS(id, "data"), filter = is.data.frame),
    selectVarInput(NS(id, "var"))
  )
}
selectDataVarServer <- function(id, filter = is.numeric) {
  moduleServer(id, function(input, output, session) {
    data <- datasetServer("data")
    var <- selectVarServer("var", data, filter = filter)
    var
  })
}

selectDataVarApp <- function(filter = is.numeric) {
  ui <- fluidPage(
    sidebarLayout(
      sidebarPanel(selectDataVarUI("var")),
      mainPanel(verbatimTextOutput("out"))
```

```
    )
  )
  server <- function(input, output, session) {
    var <- selectDataVarServer("var", filter)
    output$out <- renderPrint(var(), width = 40)
  }
  shinyApp(ui, server)
}
```

Case Study: Histogram

Now let's circle back to the original histogram module and refactor it into something more composable. The key challenge of creating modules is creating functions that are flexible enough to be used in multiple places but simple enough that they can be easily understood. Figuring out how to write functions that are good building blocks is the journey of a lifetime; expect that you'll have to do it wrong quite a few times before you get it right. (I wish I could offer more concrete advice here, but currently this is a skill that you'll have to refine through practice and conscious reflection.)

I'm also going to consider it as an output control because while it does use an input (the number of bins), that's used only to tweak the display and doesn't need to be returned by the module:

```
histogramOutput <- function(id) {
  tagList(
    numericInput(NS(id, "bins"), "bins", 10, min = 1, step = 1),
    plotOutput(NS(id, "hist"))
  )
}
```

I've decided to give this module two inputs: x, the variable to plot, and a title for the histogram. Both will be reactive so that they can change over time. (The title is a bit frivolous, but it's going to motivate an important technique very shortly.) Note the default value of title—it has to be reactive, so we need to wrap a constant value inside of reactive():

```
histogramServer <- function(id, x, title = reactive("Histogram")) {
  stopifnot(is.reactive(x))
  stopifnot(is.reactive(title))

  moduleServer(id, function(input, output, session) {
    output$hist <- renderPlot({
      req(is.numeric(x()))
      main <- paste0(title(), " [", input$bins, "]")
      hist(x(), breaks = input$bins, main = main)
    }, res = 96)
  })
}
```

```
histogramApp <- function() {
  ui <- fluidPage(
    sidebarLayout(
      sidebarPanel(
        datasetInput("data", is.data.frame),
        selectVarInput("var"),
      ),
      mainPanel(
        histogramOutput("hist")
      )
    )
  )

  server <- function(input, output, session) {
    data <- datasetServer("data")
    x <- selectVarServer("var", data)
    histogramServer("hist", x)
  }
  shinyApp(ui, server)
}
# histogramApp()
```

 Note that if you wanted to allow the module user to place the breaks control and histogram in different places of the app, you could use multiple UI functions. It's not terribly useful here, but it's useful to see the basic approach:

```
histogramOutputBins <- function(id) {
  numericInput(NS(id, "bins"), "bins", 10, min = 1, step = 1)
}
histogramOutputPlot <- function(id) {
  plotOutput(NS(id, "hist"))
}

ui <- fluidPage(
  sidebarLayout(
    sidebarPanel(
      datasetInput("data", is.data.frame),
      selectVarInput("var"),
      histogramOutputBins("hist")
    ),
    mainPanel(
      histogramOutputPlot("hist")
    )
  )
)
```

Multiple Outputs

It would be nice if we could include the name of the selected variable in the title of the histogram. There's currently no way to do that because selectVarServer() only returns the value of the variable, not its name. We could certainly rewrite `selectVar Server()` to return the name instead, but then the module user would have to do the subsetting. A better approach would be for the `selectVarServer()` to return *both* the name and the value.

A server function can return multiple values exactly the same way that any R function can return multiple values: by returning a list. Here, we modify `selectVarServer()` to return both the name and value as reactives:

```r
selectVarServer <- function(id, data, filter = is.numeric) {
  stopifnot(is.reactive(data))
  stopifnot(!is.reactive(filter))

  moduleServer(id, function(input, output, session) {
    observeEvent(data(), {
      updateSelectInput(session, "var", choices = find_vars(data(), filter))
    })

    list(
      name = reactive(input$var),
      value = reactive(data()[[input$var]])
    )
  })
}
```

Now we can update our `histogramApp()` to make use of this. The UI stays the same; but now we pass both the selected variable's value and its name to `histogram Server()`:

```r
histogramApp <- function() {
  ui <- fluidPage(...)

  server <- function(input, output, session) {
    data <- datasetServer("data")
    x <- selectVarServer("var", data)
    histogramServer("hist", x$value, x$name)
  }
  shinyApp(ui, server)
}
```

The main challenge with this sort of code is remembering when you use the reactive (e.g., `x$value`) versus when you use its value (e.g., `x$value()`). Just remember that when passing an argument to a module, you want the module to react to the value changing, which means that you have to pass the reactive, not its current value.

If you find yourself frequently returning multiple values from a reactive, you might also consider using the zeallot (*https://github.com/r-lib/zeallot*) package. zeallot provides the `%<-%` operator, which allows you to assign into multiple variables (sometimes called multiple, unpacking, or destructuring assignment). This can be useful when returning multiple values because you avoid a layer of indirection:

```
library(zeallot)

histogramApp <- function() {
  ui <- fluidPage(...)

  server <- function(input, output, session) {
    data <- datasetServer("data")
    c(value, name) %<-% selectVarServer("var", data)
    histogramServer("hist", value, name)
  }
  shinyApp(ui, server)
}
```

Exercises

1. Rewrite selectVarServer() so that both data and filter are reactive. Then use it with an app function that lets the user pick the dataset with the dataset module and filtering function using inputSelect(). Give the user the ability to filter numeric, character, or factor variables.

2. The following code defines output and server components of a module that takes a numeric input and produces a bulleted list of three summary statistics. Create an app function that allows you to experiment with it. The app function should take a data frame as input and use numericVarSelectInput() to pick the variable to summarize:

```
summaryOutput <- function(id) {
  tags$ul(
    tags$li("Min: ", textOutput(NS(id, "min"), inline = TRUE)),
    tags$li("Max: ", textOutput(NS(id, "max"), inline = TRUE)),
    tags$li("Missing: ", textOutput(NS(id, "n_na"), inline = TRUE))
  )
}

summaryServer <- function(id, var) {
  moduleServer(id, function(input, output, session) {
    rng <- reactive({
      req(var())
      range(var(), na.rm = TRUE)
    })

    output$min <- renderText(rng()[[1]])
    output$max <- renderText(rng()[[2]])
    output$n_na <- renderText(sum(is.na(var())))
  })
}
```

3. The following module input provides a text control that lets you type a date in ISO8601 format (yyyy-mm-dd). Complete the module by providing a server

function that uses `output$error` to display a message if the entered value is not a valid date. The module should return a `Date` object for valid dates. (Hint: use `strptime(x, "%Y-%m-%d")` to parse the string; it will return `NA` if the value isn't a valid date.)

```
ymdDateUI <- function(id, label) {
  label <- paste0(label, " (yyyy-mm-dd)")

  fluidRow(
    textInput(NS(id, "date"), label),
    textOutput(NS(id, "error"))
  )
}
```

Case Studies

To summarize what you've learned so far:

- Module inputs (i.e., additional arguments to the module server) can be reactives or constants. The choice is a design decision that you make based on who sets the arguments and when they change. You should always check that the arguments are of the expected type to avoid unhelpful error messages.

- Unlike app servers, but like regular functions, module servers can return values. The return value of a module should always be a reactive or, if you want to return multiple values, a list of reactives.

To help these ideas to sink in, I'll present a few case studies that show a few more examples of using modules. Unfortunately, I don't have the space to show every possible way you might use a module to help simplify your app, but hopefully these examples will give you a little flavor for what you can do and suggest directions to consider in the future.

Limited Selection and Other

Another important use of modules is to give complex UI elements a simpler user interface. Here I'm going to create a useful control that Shiny doesn't provide by default: a small set of options displayed with radio buttons coupled with an "Other" field. The inside of this module uses multiple input elements, but from the outside it works as a single combined object.

I'm going to parameterize the UI side with `label`, `choices`, and `selected`, which get passed directly to `radioButtons()`. I also create a `textInput()` containing a placeholder, which defaults to "Other." To combine the text box and the radio button, I take advantage of the fact that `choiceNames` can be a list of HTML elements, including other input widgets. Figure 19-3 gives you a sense of what it'll look like:

```r
radioExtraUI <- function(id, label, choices, selected = NULL, placeholder = "Other") {
  other <- textInput(NS(id, "other"), label = NULL, placeholder = placeholder)

  names <- if (is.null(names(choices))) choices else names(choices)
  values <- unname(choices)

  radioButtons(NS(id, "primary"),
    label = label,
    choiceValues = c(names, "other"),
    choiceNames = c(as.list(values), list(other)),
    selected = selected
  )
}
```

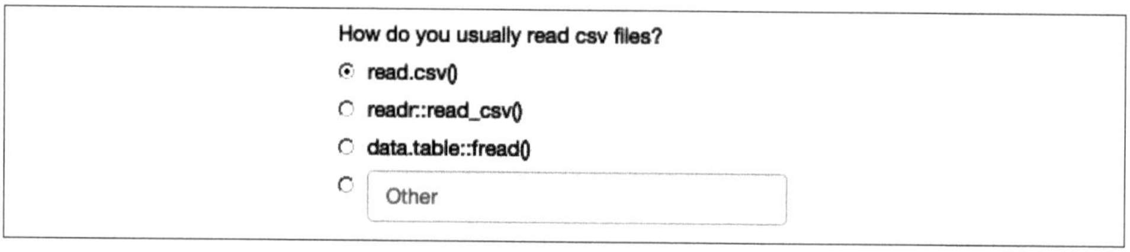

Figure 19-3. An example using `radioExtraUI()` *to find out how you usually read CSV files.*

On the server, I want to automatically select the "Other" radio button if you modify the placeholder value. You could also imagine using validation to ensure that some text is present if Other is selected:

```r
radioExtraServer <- function(id) {
  moduleServer(id, function(input, output, session) {
    observeEvent(input$other, ignoreInit = TRUE, {
      updateRadioButtons(session, "primary", selected = "other")
    })

    reactive({
      if (input$primary == "other") {
        input$other
      } else {
        input$primary
      }
    })
  })
}
```

Then I wrap up both pieces in an app function so that I can test it. Here I use ... to pass down any number of arguments into my `radioExtraUI()`:

```r
radioExtraApp <- function(...) {
  ui <- fluidPage(
    radioExtraUI("extra", ...),
    textOutput("value")
  )
```

```
server <- function(input, output, server) {
  extra <- radioExtraServer("extra")
  output$value <- renderText(paste0("Selected: ", extra()))
}

shinyApp(ui, server)
}
```

Figure 19-4 gives you a sense of how it behaves:

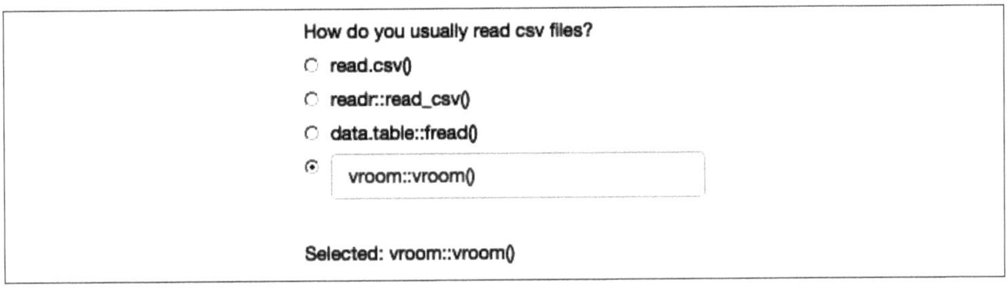

> **How do you usually read csv files?**
>
> ○ read.csv()
>
> ○ readr::read_csv()
>
> ○ data.table::fread()
>
> ◉ [vroom::vroom()]
>
> Selected: vroom::vroom()

Figure 19-4. Testing `radioExtraApp()` with the same question about how you read CSVs. Now, if you type something in the other field, the corresponding radio button is automatically selected.

You could continue to wrap up this module for still more specific purposes. For example, one variable that requires a little care is gender, because there are many different ways for people to express their gender:

```
genderUI <- function(id, label = "Gender") {
  radioExtraUI(id,
    label = label,
    choices = c(
      male = "Male",
      female = "Female",
      na = "Prefer not to say"
    ),
    placeholder = "Self-described",
    selected = "na"
  )
}
```

Here it's important to provide the most common choices, male and female; an option to not provide that data; and then a write-in option where people can use whatever term they're most comfortable with. It's considerate not to use a placeholder of "Other" here.

For a deeper dive into this issue, and a discussion of why many commonly used ways of asking about gender can be hurtful to some people, I recommend reading "Designing Forms for Gender Diversity and Inclusion" (*https://uxdesign.cc/d8194cf1f51*) by Sabrina Fonseca or "Standard for Sex, Gender, Variations of Sex Characteristics and

Sexual Orientation Variables" (*https://oreil.ly/D9snI*) by the Australian Bureau of Statistics.

Wizard

Next we'll tackle a pair of case studies that dive into some subtleties of namespacing, where the UI is generated at different times by different people. These situations are complex because you need to remember the details of how namespacing works.

We'll start with a module that wraps up a wizard interface, a style of UI where you break a complex process down into a series of simple pages that the user works through one by one. I showed how to create a basic wizard in "Wizard Interface" on page 165. Now we'll automate the process so that when creating a wizard you can focus on the content of each page rather than on how they are connected together to form a whole.

To explain this module, I'm going to start from the bottom and work my way up. The main part of the wizard UI are the buttons. Each page has two buttons: one to take them to the next page and one to return them to the previous page. We'll start by creating helpers to build these buttons:

```
nextPage <- function(id, i) {
  actionButton(NS(id, paste0("go_", i, "_", i + 1)), "next")
}
prevPage <- function(id, i) {
  actionButton(NS(id, paste0("go_", i, "_", i - 1)), "prev")
}
```

The only real complexity here is the id: since each input element needs to have a unique ID, the ID for each button needs to include both the current and the destination page.

Next I write a function to generate a page of the wizard. This includes a "title" (not shown but used to identify the page for switching), the contents of the page (supplied by the user), and the two buttons:[3]

```
wrapPage <- function(title, page, button_left = NULL, button_right = NULL) {
  tabPanel(
    title = title,
    fluidRow(
      column(12, page)
    ),
    fluidRow(
      column(6, button_left),
      column(6, button_right)
```

---

3 Not every page will have both buttons (more on that shortly), so I mark them as optional by supplying a default value of NULL.

```
      )
    )
  }
```

Then we can put it all together to generate the whole wizard (Figure 19-5). We loop over the list of pages provided by the user, create the buttons, wrap up the user-supplied page into a tabPanel, then combine all the panels into a tabsetPanel. Note that there are two special cases for buttons:

- The first page doesn't have a previous button. Here I use a trick that if returns NULL if the condition is FALSE and there is no else block.

- The last page uses an input control supplied by the user. I think this is the simplest way to allow the user to control what happens when the wizard is done.

```
wizardUI <- function(id, pages, doneButton = NULL) {
  stopifnot(is.list(pages))
  n <- length(pages)

  wrapped <- vector("list", n)
  for (i in seq_along(pages)) {
    # First page only has next; last page only prev + done
    lhs <- if (i > 1) prevPage(id, i)
    rhs <- if (i < n) nextPage(id, i) else doneButton
    wrapped[[i]] <- wrapPage(paste0("page_", i), pages[[i]], lhs, rhs)
  }

  # Create tabsetPanel
  # https://github.com/rstudio/shiny/issues/2927
  wrapped$id <- NS(id, "wizard")
  wrapped$type <- "hidden"
  do.call("tabsetPanel", wrapped)
}
```

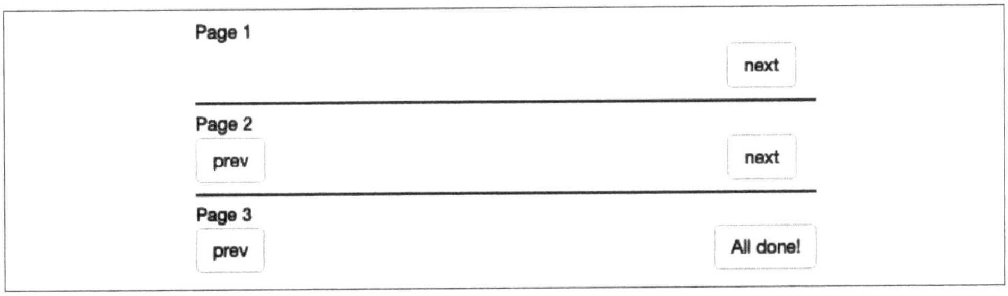

Figure 19-5. A simple example of the wizard UI.

The code to create the tabset panel requires a little explanation: unfortunately, tabset Panel() doesn't allow us to pass in a list of tabs. So instead we need to do a little do.call() magic to make it work. do.call(function_name, list(arg1, arg2, …) is equivalent to function_name(arg1, arg2, …), so here we're creating a call like tabsetPanel(pages[[1]], pages[[2]], …, id = NS(id, "wizard"), type = "hidden"). Hopefully this will be simplified in a future version of Shiny.

Now that we've completed the module UI, we need to turn our attention to the module server. The essence of the server is straightforward: we just need to make buttons work so that you can travel from page to page in either direction. To do that we need to set up an observeEvent() for each button that calls updateTabsetPanel(). This would be relatively simple if we knew exactly how many pages there were. But we don't because the user of the module gets to control that.

So instead we need to do a little functional programming to set up the (n - 1) * 2 observers (two observers for each page except for the first and last, which only need one). The following server function starts by extracting out the basic code we need for one button in the changePage() function. It uses input[[]], as in "Multiple Controls" on page 168, so we can refer to control dynamically. Then we use lapply() to loop over all the previous buttons (needed for every page except the first) and all the next buttons (needed for every page except the last):

```
wizardServer <- function(id, n) {
  moduleServer(id, function(input, output, session) {
    changePage <- function(from, to) {
      observeEvent(input[[paste0("go_", from, "_", to)]], {
        updateTabsetPanel(session, "wizard", selected = paste0("page_", to))
      })
    }
    ids <- seq_len(n)
    lapply(ids[-1], function(i) changePage(i, i - 1))
    lapply(ids[-n], function(i) changePage(i, i + 1))
  })
}
```

Note that it's not possible to use a for loop instead of map()/lapply() here. A for loop works by changing the value of the same i variable so that by the time the loop is done, every changePage() would use the same value. map() and lapply() work by creating new environments, each with their own value of i.

Now we can construct an app and simple example to make sure we've plumbed everything together correctly:

```
wizardApp <- function(...) {
  pages <- list(...)

  ui <- fluidPage(
    wizardUI("whiz", pages)
  )
  server <- function(input, output, session) {
    wizardServer("whiz", length(pages))
  }
```

```
  shinyApp(ui, server)
}
```

Unfortunately, we need to repeat ourselves slightly when using the module, and we need to make sure that the n argument to wizardServer() is consistent with the pages argument to wizardUi(). This is a principled limitation of the module system, which we'll discuss in more detail in "Single Object Modules" on page 284.

Now let's use the wizard in a slightly more realistic app that has inputs and outputs and yields Figure 19-6. The main point to notice is that even though the pages are displayed by the module, their IDs are controlled by the user of the module. The developer who creates the component controls the name; it doesn't matter who assembles the control for final display on the webpage:

```
page1 <- tagList(
  textInput("name", "What's your name?")
)
page2 <- tagList(
  numericInput("age", "How old are you?", 20)
)
page3 <- tagList(
  "Is this data correct?",
  verbatimTextOutput("info")
)

ui <- fluidPage(
  wizardUI(
    id = "demographics",
    pages = list(page1, page2, page3),
    doneButton = actionButton("done", "Submit")
  )
)
server <- function(input, output, session) {
  wizardServer("demographics", 3)

  observeEvent(input$done, showModal(
    modalDialog("Thank you!", footer = NULL)
  ))

  output$info <- renderText(paste0(
    "Age: ", input$age, "\n",
    "Name: ", input$name, "\n"
  ))
}
```

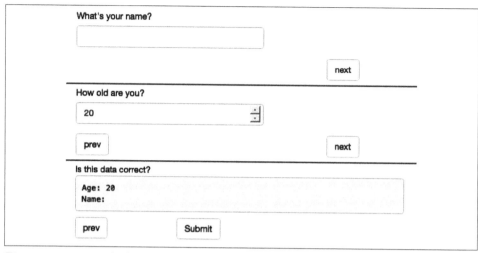

Figure 19-6. A simple, but complete, wizard created with our new module.

Dynamic UI

We'll finish up with a case study that uses dynamic UI, taking part of the dynamic filtering code found in "Dynamic Filtering" on page 171 and turning it into a module. The main challenge of dynamic UI within a module is that because you will be generating UI code within your server function, we need a more precise definition of when explicit namespacing is needed.

As usual, we'll start with the module UI. It's very simple here, because we're just generating a "hole" that the server functions will fill in dynamically:

```
filterUI <- function(id) {
  uiOutput(NS(id, "controls"))
}
```

To create the module server, we'll first copy in the helper functions from "Dynamic Filtering" on page 171: make_ui() makes a control for each column, and then fil ter_var() helps generate the final logical vector. There's only one difference here: make_ui() gains an additional id argument so that we can namespace the controls to the module:

```
library(purrr)

make_ui <- function(x, id, var) {
  if (is.numeric(x)) {
    rng <- range(x, na.rm = TRUE)
    sliderInput(id, var, min = rng[1], max = rng[2], value = rng)
  } else if (is.factor(x)) {
    levs <- levels(x)
    selectInput(id, var, choices = levs, selected = levs, multiple = TRUE)
```

```
  } else {
    # Not supported
    NULL
  }
}
filter_var <- function(x, val) {
  if (is.numeric(x)) {
    !is.na(x) & x >= val[1] & x <= val[2]
  } else if (is.factor(x)) {
    x %in% val
  } else {
    # No control, so don't filter
    TRUE
  }
}
```

Now we create the module server. There are two main parts:

- We generate the controls using purrr::map() and make_ui(). Note the explicit use of NS() here. That's needed because even though we're inside the module server, the automatic namespacing only applies to input, output, and session.

- We return the logical filtering vector as the module output.

```
filterServer <- function(id, df) {
  stopifnot(is.reactive(df))

  moduleServer(id, function(input, output, session) {
    vars <- reactive(names(df()))

    output$controls <- renderUI({
      map(vars(), function(var) make_ui(df()[[var]], NS(id, var), var))
    })

    reactive({
      each_var <- map(vars(), function(var) filter_var(df()[[var]], input[[var]]))
      reduce(each_var, `&`)
    })
  })
}
```

Now we can put it all together in a module app that allows you to select a built-in dataset and then filter on any numeric or categorical variable:

```
filterApp <- function() {
  ui <- fluidPage(
    sidebarLayout(
      sidebarPanel(
        datasetInput("data", is.data.frame),
        textOutput("n"),
        filterUI("filter"),
      ),
      mainPanel(
```

```
      tableOutput("table")
    )
  )
)
server <- function(input, output, session) {
  df <- datasetServer("data")
  filter <- filterServer("filter", df)

  output$table <- renderTable(df()[filter(), , drop = FALSE])
  output$n <- renderText(paste0(sum(filter()), " rows"))
}
shinyApp(ui, server)
}
```

A big advantage of using a module here is that it wraps up a bunch of advanced Shiny programming techniques. You can use the filter module without having to understand the dynamic UI and functional programming techniques that make it work.

Single Object Modules

To conclude the chapter, I wanted to finish up with a brief discussion of a common reaction to modules. Feel free to skip this section if that wasn't your reaction. When some people (like me!) encounter modules for the first time, they immediately attempt to combine the module server and module UI into a single-module object. To illustrate the problem, let's generalize the motivating example from the first part of the chapter so that the data frame is now a parameter:

```
histogramUI <- function(id, df) {
  tagList(
    selectInput(NS(id, "var"), "Variable", names(df)),
    numericInput(NS(id, "bins"), "bins", 10, min = 1),
    plotOutput(NS(id, "hist"))
  )
}

histogramServer <- function(id, df) {
  moduleServer(id, function(input, output, session) {
    data <- reactive(df[[input$var]])
    output$hist <- renderPlot({
      hist(data(), breaks = input$bins, main = input$var)
    }, res = 96)
  })
}
```

And that leads to the following app:

```
ui <- fluidPage(
  tabsetPanel(
    tabPanel("mtcars", histogramUI("mtcars", mtcars)),
    tabPanel("iris", histogramUI("iris", iris))
  )
```

```
    )
    server <- function(input, output, session) {
      histogramServer("mtcars", mtcars)
      histogramServer("iris", iris)
    }
```

It seems undesirable that we have to repeat both the ID and the name of the dataset in both the UI and server, so it's natural to want to wrap into a single function that returns both the UI and the server:

```
    histogramApp <- function(id, df) {
      list(
        ui = histogramUI(id, df),
        server = histogramServer(id, df)
      )
    }
```

Then we define the module outside of the UI and server, extracting elements from the list as needed:

```
    hist1 <- histogramApp("mtcars", mtcars)
    hist2 <- histogramApp("iris", iris)

    ui <- fluidPage(
      tabsetPanel(
        tabPanel("mtcars", hist1$ui()),
        tabPanel("iris", hist2$ui())
      )
    )
    server <- function(input, output, session) {
      hist1$server()
      hist2$server()
    }
```

There are two problems with this code. First, it doesn't work, because module Server() must be called inside a server function. But imagine that problem didn't exist or you worked around it some other way. There's still a big problem: what if we want to allow the user to select the dataset (i.e., we want to make the df argument reactive)? That can't work because the module is instantiated before the server function (i.e., before we know that information).

In Shiny, UI and server are inherently disconnected; Shiny doesn't know which UI invocation belongs to which server session. You can see this pattern throughout Shiny: for example, plotOutput() and renderPlot() are connected only by shared ID. Writing modules as separate functions reflects that reality: they're distinct functions that are not connected other than through a shared ID.

Summary

This chapter has shown you how to use Shiny modules, a generalization of functions that allow you to extract coordinated UI and server code into reusable components. It takes a while to get your head around modules, but once you do, you'll have unlocked a powerful technique for simplifying complex apps.

In the next chapter, you'll learn how to structure your Shiny app like a package so that you can take advantage of the testing tools available to R packages.

## CHAPTER 20

Packages

If you are creating a large or long-term Shiny app, I highly recommend that you organize your app in the same way as an R package. This means that you:

- Put all R code in the *R/* directory.
- Write a function that starts your app (i.e., that calls `shinyApp()` with your UI and server).
- Create a `DESCRIPTION` file in the root directory of your app.

This structure gets your toes into the water of package development. It's a long way from a complete package, but it's still useful because it activates new tools that make it easier to work with a larger app. The package structure will pay off further when we talk about testing in Chapter 21, because you get tools that make it easy to run the tests and to see what code is tested. In the long run, it also helps you document complex apps using roxygen2 (*https://roxygen2.r-lib.org*), although we won't discuss that in this book.

It's easy to think of packages as giant complicated things like Shiny, ggplot2, or dplyr. But packages can also be very simple. The core idea of a package is that it's a set of conventions for organizing your code and related artifacts: if you follow those conventions, you get a bunch of tools for free. In this chapter, I'll show you the most important conventions and then provide a few hints as to next steps.

As you start working with app-packages, you may find that you enjoy the process of package development and want to learn more. I'd suggest starting with *R Packages* (*https://r-pkgs.org*) to get the lay of the package development land, then continuing on to *Engineering Production Grade Shiny Apps* (*http://engineering-shiny.org*), by Colin Fay, Sébastien Rochette, Vincent Guyader, and Cervan Girard to learn more about the intersection of R packages and Shiny apps.

As usual, we begin by loading shiny:

```
library(shiny)
```

Converting an Existing App

Converting an app to a package requires some up-front work. Assuming that you have an app called myApp and it already lives in a directory called *myApp/*, you'll need to do the following things:

- Create an *R* directory and move *app.R* into it.
- Transform your app into a standalone function by wrapping:

  ```
  library(shiny)

  myApp <- function(...) {
    ui <- fluidPage(
      ...
    )
    server <- function(input, output, session) {
      ...
    }
    shinyApp(ui, server, ...)
  }
  ```

- Call usethis::use_description() to create a description file. In many cases, you'll never need to look at this file, but you need it to activate RStudio's "package development mode," which provides the keyboard shortcuts we'll use later.
- If you don't already have one, create an RStudio project by calling use this::use_rstudio().
- Restart RStudio and reopen your project.

You can now press Cmd/Ctrl+Shift+L to run devtools::load_all() and load all the package code and data. This means that you can now:

- Remove any calls to source(), since load_all() automatically sources all *.R* files in *R/*.
- If you are loading datasets using read.csv() or similar, you can instead use use this::use_data(mydataset) to save the data in the *data/* directory. load_all() automatically loads the data for you.

To make this process more concrete, we'll next work through a simple case study before coming back to the other benefits of this work in "Benefits" on page 292.

Single File

Imagine I have a relatively complex app that currently lives in a single *app.R*:

```r
library(shiny)

monthFeedbackUI <- function(id) {
  textOutput(NS(id, "feedback"))
}
monthFeedbackServer <- function(id, month) {
  stopifnot(is.reactive(month))

  moduleServer(id, function(input, output, session) {
    output$feedback <- renderText({
      if (month() == "October") {
        "You picked a great month!"
      } else {
        "Eh, you could do better."
      }
    })
  })
}

stones <- vroom::vroom("birthstones.csv")
birthstoneUI <- function(id) {
  p(
    "The birthstone for ", textOutput(NS(id, "month"), inline = TRUE),
    " is ", textOutput(NS(id, "stone"), inline = TRUE)
  )
}
birthstoneServer <- function(id, month) {
  stopifnot(is.reactive(month))

  moduleServer(id, function(input, output, session) {
    stone <- reactive(stones$stone[stones$month == month()])
    output$month <- renderText(month())
    output$stone <- renderText(stone())
  })
}

months <- c(
  "January", "February", "March", "April", "May", "June",
  "July", "August", "September", "October", "November", "December"
)
ui <- navbarPage(
  "Sample app",
  tabPanel("Pick a month",
    selectInput("month", "What's your favourite month?", choices = months)
  ),
  tabPanel("Feedback", monthFeedbackUI("tab1")),
  tabPanel("Birthstone", birthstoneUI("tab2"))
)
```

```
server <- function(input, output, session) {
  monthFeedbackServer("tab1", reactive(input$month))
  birthstoneServer("tab2", reactive(input$month))
}
shinyApp(ui, server)
```

This code creates a simple three-page app that uses modules to keep the pages isolated. It's a toy app, but it's still realistic. The main difference compared to a real app is that here the individual UI and server components are much simpler.

Module Files

Before turning it into a package, my first step is to pull the two modules out into their own files following the advice in "Naming Conventions" on page 265:

- *R/monthFeedback.R*:

  ```
  monthFeedbackUI <- function(id) {
    textOutput(NS(id, "feedback"))
  }
  monthFeedbackServer <- function(id, month) {
    stopifnot(is.reactive(month))

    moduleServer(id, function(input, output, session) {
      output$feedback <- renderText({
        if (month() == "October") {
          "You picked a great month!"
        } else {
          "Eh, you could do better."
        }
      })
    })
  }
  ```

- *R/birthstone.R*:

  ```
  birthstoneUI <- function(id) {
    p(
      "The birthstone for ", textOutput(NS(id, "month"), inline = TRUE),
      " is ", textOutput(NS(id, "stone"), inline = TRUE)
    )
  }
  birthstoneServer <- function(id, month) {
    stopifnot(is.reactive(month))

    moduleServer(id, function(input, output, session) {
      stone <- reactive(stones$stone[stones$month == month()])
      output$month <- renderText(month())
      output$stone <- renderText(stone())
  ```

```
    })
  }
```

That leaves me with the following *app.R*:

```r
library(shiny)

stones <- vroom::vroom("birthstones.csv")
#> Rows: 12
#> Columns: 2
#> Delimiter: ","
#> chr [2]: month, stone
#>
#> Use `spec()` to retrieve the guessed column specification
#> Pass a specification to the `col_types` argument to quiet this message
months <- c(
  "January", "February", "March", "April", "May", "June",
  "July", "August", "September", "October", "November", "December"
)

ui <- navbarPage(
  "Sample app",
  tabPanel("Pick a month",
    selectInput("month", "What's your favourite month?", choices = months)
  ),
  tabPanel("Feedback", monthFeedbackUI("tab1")),
  tabPanel("Birthstone", birthstoneUI("tab2"))
)
server <- function(input, output, session) {
  monthFeedbackServer("tab1", reactive(input$month))
  birthstoneServer("tab2", reactive(input$month))
}
shinyApp(ui, server)
```

Just pulling the modules out into separate files is useful because it helps me understand the big picture of the app. If instead I want to dive into the details, I can look at the modules files.

A Package

Now let's make this into a package. First I run usethis::use_description(), which creates a DESCRIPTION file. Next, I move *app.R* to *R/app.R* and wrap shinyApp() into a function:

```r
library(shiny)

monthApp <- function(...) {
  stones <- vroom::vroom("birthstones.csv")
  months <- c(
    "January", "February", "March", "April", "May", "June",
    "July", "August", "September", "October", "November", "December"
```

```
  )

  ui <- navbarPage(
    "Sample app",
    tabPanel("Pick a month",
      selectInput("month", "What's your favourite month?", choices = months)
    ),
    tabPanel("Feedback", monthFeedbackUI("tab1")),
    tabPanel("Birthstone", birthstoneUI("tab2"))
  )
  server <- function(input, output, session) {
    monthFeedbackServer("tab1", reactive(input$month))
    birthstoneServer("tab2", reactive(input$month))
  }
  shinyApp(ui, server, ...)
}
```

As an optional extra, I converted *birthstones.csv* to a package dataset by running `use this::use_data("birthstones")`. This creates *data/birthstones.rda*, which will be loaded automatically when I load the package. I can now delete *birthstones.csv* and remove the line that reads it in: `stones <- vroom::vroom("birthstones.csv")`.

You can see the final product on GitHub (*https://github.com/hadley/monthApp*).

Benefits

Why bother doing all this work? The most important benefit is a new workflow that makes it easier to accurately reload all app code and relaunch the app. But it also makes it easier to share code between apps and share your app with others.

Workflow

Putting your app code into the package structure unlocks a new workflow:

- Reload all code in the app with Cmd/Ctrl+Shift+L. This calls `dev tools::load_all()`, which automatically saves all open files, `source()`s every file in *R/*, loads all datasets in *data/*, then puts your cursor in the console.

- Rerun the app with `myApp()`.

As your app grows bigger, it's also worth knowing about the two most important code navigation keyboard shortcuts:

- Ctrl/Cmd+. (period) will open the "fuzzy file and function finder." Type a few letters at the start of the file or function that you want to navigate to, select it with the arrow keys, and then press Enter. This allows you to quickly jump around your app without taking your hands off the keyboard.

- When your cursor is on the name of the function, F2 will jump to the function definition.

If you do a lot of package development, you might want to automatically load use this, so you can type (for example) `use_description()` instead of `use this::use_description()`. You can do so by adding the following lines to your *.Rprofile*. This file contains R code that's run whenever you start R, so it's a great way to customize your interactive development environment:

```
if (interactive()) {
  require(usethis, quietly = TRUE)
}
```

The easiest way to edit this file is to call `usethis::edit_r_profile()`.

Sharing

Since your app is now wrapped up in a function, it's easy to include multiple apps in the same package. And because you have multiple apps in the same place, it's now much easier to share code and data across apps. That's a huge benefit if you have a bunch of apps for related tasks.

Packages are also a great way to share apps. shinyapps.io (*https://www.shinyapps.io*) and RStudio Connect (*https://rstudio.com/products/connect*) are useful for sharing apps with folks who aren't familiar with R. But sometimes you want to share apps with your colleagues who do use R. Maybe instead of allowing the user to upload a dataset, you want to provide them with a function that they call with a data frame. For example, the following very simple app allows the R user to supply their own data frame for interactive summaries:

```
dataSummaryApp <- function(df) {
  ui <- fluidPage(
    selectInput("var", "Variable", choices = names(df)),
    verbatimTextOutput("summary")
  )

  server <- function(input, output, session) {
    output$summary <- renderPrint({
      summary(df[[input$var]])
    })
  }

  shinyApp(ui, server)
}
```

RStudio Gadgets (*https://oreil.ly/XhwV9*) build on this idea: they are Shiny apps that let you add a new user interface to the RStudio IDE. It's even possible to write gadgets that generate code, so you can perform some task that's easy to do interactively, and the gadget generates the corresponding code and saves back into the open file.

Extra Steps

There are two common extra steps you might take beyond the basics: making it easy to deploy your app-package and turning it into a "real" package.

Deploying Your App-Package

If you want to deploy your app to RStudio Connect or Shiny,[1] you'll need two extra steps:

- You'll need an *app.R* that tells the deployment server how to run your app. The easiest way is to load the code with pkgload:

  ```
  pkgload::load_all(".")
  myApp()
  ```

  You can see other techniques in Chapter 13 of *Engineering Shiny* (*https://engineering-shiny.org/deploy.html*).

- Normally when you deploy an app, the rsconnect package automatically figures out all of the packages your code uses. But now that you have a DESCRIPTION file, it requires you to explicitly specify them. The easiest way to do this is to call use this::use_package(). You'll need to start with shiny and pkgload:

  ```
  usethis::use_package("shiny")
  usethis::use_package("pkgload")
  ```

  This is a little more work, but the payoff is a having an explicit list of every package that your app needs in one place.

Now you can run rsconnect::deployApp() whenever you're ready to share an updated version of your app with your users.

R CMD check

A minimal package contains an *R/* directory, a DESCRIPTION file, and a function to run your app. As you've seen, this is already useful because it unlocks some workflows to speed up app development. But what makes a "real" app? To me, it's making a serious effort to get R CMD check passing. R CMD check is R's automated system that checks your package for common problems. In RStudio, you can run R CMD check by pressing Cmd/Ctrl+Shift+E.

I don't recommend that you do this the first time, the second time, or even the third time you try out the package structure. Instead, I recommend that you get familiar

---

[1] I'd expect most other ways of deploying Shiny apps would also work since *app.R* is the most common way of structuring apps.

with the basic structure and workflow before you take the next step to make a fully compliant package. It's also something I'd generally reserve for important apps, particularly any app that will be deployed elsewhere. It can be a lot of work to get R CMD check passing, and there's little payoff in the short term. But in the long term this will protect you against a number of potential problems, and because it ensures your app adheres to standards that R developers are familiar with, it makes it easier for others to contribute to your app.

Before you make your first full app-package, you should read "The Whole Game" (*https://r-pkgs.org/whole-game.html*) chapter of *R Packages*: it will give you a fuller sense of the package structure and introduce you to other useful workflows. Then use the following hints to get R CMD check passing cleanly:

- Remove any calls to library() or require() and instead replace them with a declaration in your DESCRIPTION. Use usethis::use_package("name") to add the required package to the DESCRIPTION.[2] You'll then need to decide whether you want to refer to each function explicitly with :: or use @importFrom package Name functionName to declare the import in one place.

  At a minimum, you'll need usethis::use_package("shiny"), and for Shiny apps, I recommend using @import shiny to make all the functions in the Shiny package easily available. (Using @import is not generally considered best practice, but it makes sense here.)

- Pick a license and then use the appropriate use_license_ function to put it in the right place. For proprietary code, you can use usethis::use_propriet ary_license(). See Chapter 9 of *R Packages* (*https://r-pkgs.org/license.html*) for more details.

- Add *app.R* to *.Rbuildignore* with usethis::use_build_ignore("app.R") or similar.

- If your app contains small reference datasets, put them in data or inst/extdata. We discussed usethis::use_data() previously; alternatively, you can put raw data in inst/ext and load it with read.csv(system.file("exdata", "mydata.csv", package = "myApp")) or similar.

- You can also change your *app.R* to use the package. This requires that your package is available somewhere that your deployment machine can install from. For public work, this means a CRAN or GitHub package; for private work, this may

---

2 The distinction between Imports and Suggests is not generally important for app packages. If you do want to make a distinction, the most useful method is to use Imports for packages that need to be present on the deployment machine (in order for the app to work) and Suggests for packages that need to be present on the development machine (in order to develop the app).

mean using a tool like RStudio Package Manager (*https://oreil.ly/mOIVP*) or drat (*https://oreil.ly/binbF*):

```
myApp::myApp()
```

Summary

In this chapter, you've dipped your toes into the water of package development. This might seem overwhelming if you think of packages like ggplot2 and shiny, but packages can be very, very simple. In fact, all a project needs to be a package is a directory of R files and a DESCRIPTION file. A package is just a lightweight set of conventions that unlock useful tools and workflows. In this chapter, you learned how to turn an app into a package and some of the reasons why you might want to. In the next chapter, you'll learn about the most important reason to turn your app into a package: to make it easier to test.

CHAPTER 21
Testing

For simple apps, it's easy enough to remember how the app is supposed to work so that when you make changes to add new features, you don't accidentally break existing capabilities. However, as your app gets more complicated, it becomes impossible to hold it all in your head simultaneously. Testing is a way to capture desired behavior of your code in such a way that you can automatically verify that it keeps working the way you expect. Turning your existing informal tests into code is painful when you first do it, because you need to carefully turn every key press and mouse click into a line of code, but once done, it's tremendously faster to rerun your tests.

We'll perform automated testing with the testthat (*http://testthat.r-lib.org*) package. testthat requires turning your app into a package, but as discussed in Chapter 20, this is not too much work, and I think it pays off for other reasons.

A testthat test looks like this:

```
test_that("as.vector() strips names", {
  x <- c(a = 1, b = 2)
  expect_equal(as.vector(x), c(1, 2))
})
```

We'll come back to the details very soon, but note that a test starts by declaring the intent ("as.vector() strips names"), then uses regular R code to generate some test data. The test data is then compared to the expected result using an *expectation*, a function that starts with expect_. The first argument is some code to run, and the second argument describes the expected result: here we verify that the output of as.vector(x) equals c(1, 2).

We'll work through four levels of testing in this chapter:

- We'll start by testing nonreactive functions. This will help you learn the basic testing workflow and allow you to verify the behavior of code that you've extracted out of the server function or UI. This is exactly the same type of testing you'd do if you were writing a package, so you can find more details in the testing chapter (*https://r-pkgs.org/tests.html*) of *R Packages*.

- Next you'll learn how to test the flow of reactivity within your server function. You'll set the value of inputs and then verify that reactives and outputs have the values you expect.

- Then we'll test parts of Shiny that use JavaScript (e.g., the `update*` functions) by running the app in a background web browser. This is a high-fidelity simulation because it runs a real browser, but on the downside, the tests are slower to run, and you can no longer so easily peek inside the app.

- Finally, we'll test *app visuals* by saving screenshots of selected elements. This is necessary for testing app layout, CSS, plots, and HTML widgets, but it's fragile because screenshots can easily change for many reasons. This means that human intervention is required to confirm whether each change is OK or not, making this the most labor-intensive form of testing.

These levels of testing form a natural hierarchy because each technique provides a fuller simulation of the user experience of an app. The downside of the better simulations is that each level is slower because it has to do more, and it's more fragile because more external forces come into play. You should always strive to work at the lowest possible level so your tests are as fast and robust as possible. Over time this will also influence the way you write code: knowing what sort of code is easier to test will naturally push you toward simpler designs. Interleaved between the different levels of testing, I'll also provide advice about testing workflow and more general testing philosophy. Let's begin:

```
library(shiny)
library(testthat) # >= 3.0.0
library(shinytest)
```

Testing Functions

The easiest part of your app to test is the part that has the least to do with Shiny: the functions extracted out of your UI and server code as described in Chapter 18. We'll start by discussing how to test these nonreactive functions, showing you the basic structure of unit testing with testthat.

Basic Structure

Tests are organized into three levels:

File

All test files live in *tests/testthat*, and each test file should correspond to a code file in *R/*. For example, the code in *R/module.R* should be tested by the code in *tests/testthat/test-module.R*. Fortunately, you don't have to remember that convention: just use `usethis::use_test()` to automatically create or locate the test file corresponding to the currently open R file.

Test

Each file is broken down into tests, that is, a call to `test_that()`. A test should generally check a single property of a function. It's hard to describe exactly what this means, but a good heuristic is that you can easily describe the test in the first argument to `test_that()`.

Expectation

Each test contains one or more expectations, with functions that start with `expect_`. These define exactly what you expect code to do, whether it's returning a specific value, throwing an error, or something else. In this chapter, I'll discuss the most important expectations for Shiny apps, but you can see the full list on the testthat website (*https://oreil.ly/Ibt5O*).

The art of testing is figuring out how to write tests that clearly define the expected behavior of your function, without depending on incidental details that might change in the future.

Basic Workflow

Now that you understand the basic structure, let's dive into some examples. I'm going to start with a simple example from "Reading Uploaded Data" on page 255. Here I've extracted out some code from my server function and called it `load_file()`:

```
load_file <- function(name, path) {
  ext <- tools::file_ext(name)
  switch(ext,
    csv = vroom::vroom(path, delim = ",", col_types = list()),
    tsv = vroom::vroom(path, delim = "\t", col_types = list()),
    validate("Invalid file; Please upload a .csv or .tsv file")
  )
}
```

For the sake of this example, I'm going to pretend this code lives in *R/load.R*, so my tests for it need to live in *tests/testthat/test-load.R*. The easiest way to create that file is to run `usethis::use_test()` with *load.R*.[1]

---

[1] If you don't use RStudio, you'll need to give `use_test()` the name of the file, like `use this::use_test("load")`.

There are three main things that I want to test for this function: can it load a CSV file, can it load a TSV file, and does it give an error message for other types? To test those three things, I'll need some sample files, which I save in the session temp directory so they're automatically cleaned up after my tests are run. Then I write three expectations, two checking that the loaded file equals the original data, and one checking that I get an error:

```
test_that("load_file() handles all input types", {
  # Create sample data
  df <- tibble::tibble(x = 1, y = 2)
  path_csv <- tempfile()
  path_tsv <- tempfile()
  write.csv(df, path_csv, row.names = FALSE)
  write.table(df, path_tsv, sep = "\t", row.names = FALSE)

  expect_equal(load_file("test.csv", path_csv), df)
  expect_equal(load_file("test.tsv", path_tsv), df)
  expect_error(load_file("blah", path_csv), "Invalid file")
})
#> Test passed ✓
```

There are four ways to run this test:

- As I'm developing it, I run each line interactively at the console. When an expectation fails, it turns into an error, which I then fix.

- Once I've finished developing it, I run the whole test block. If the test passes, I get a message like Test passed 😃. If it fails, I get the details of what went wrong.

- As I develop more tests, I run all of the tests for the current file[2] with dev tools::test_file(). Because I do this so often, I have a special keyboard shortcut set up to make it as easy as possible. I'll show you how to set that up yourself very shortly.

- Every now and then I run all of the tests for the whole package with dev tools::test(). This ensures that I haven't accidentally broken anything outside of the current file.

Key Expectations

There are two expectations that you'll use a lot of the time when testing functions: expect_equal() and expect_error(). Like all expectation functions, the first argument is the code to check, and the second argument is the expected outcome: an expected value in the case of expect_equal(), and an expected error text in the case of expect_error().

---

2 Like usethis::use_test(), this only works if you're using RStudio.

To get a sense for how these functions work, it's useful to call them directly, outside of tests.

When using expect_equal(), remember that you don't have to test that whole object: generally it's better to test just the component that you're interested in:

```
complicated_object <- list(
  x = list(mtcars, iris),
  y = 10
)
expect_equal(complicated_object$y, 10)
```

There are a few expectations for special cases of expect_equal() that can save you a little typing:

- expect_true(x) and expect_false(x) are equivalent to expect_equal(x, FALSE) and expect_equal(x, TRUE).
- expect_null(x) is equivalent to expect_equal(x, NULL).
- expect_named(x, c("a", "b", "c")) is equivalent to expect_equal(names(x), c("a", "b", "c")) but has the options ignore.order and ignore.case.
- expect_length(x, 10) is equivalent to expect_equal(length(x), 10).

There are also functions that implement relaxed versions of expect_equal() for vectors:

- expect_setequal(x, y) tests that every value in x occurs in y and every value in y occurs in x.
- expect_mapequal(x, y) tests that x and y have the same names and that x[names(y)] equals y.

It's often important to test that code generates an error, for which you can use expect_error():

```
expect_error("Hi!")
#> Error: "Hi!" did not throw the expected error.
expect_error(stop("Bye"))
```

Note that the second argument to expect_error() is a regular expression—the goal is to find a short fragment of text that matches the error you expect and is unlikely to match errors that you don't expect:

```
f <- function() {
  stop("Calculation failed [location 1]")
}

expect_error(f(), "Calculation failed [location 1]")
```

```
#> Error in f(): Calculation failed [location 1]
expect_error(f(), "Calculation failed \\[location 1\\]")
```

But it's better still to just pick a small fragment to match:

```
expect_error(f(),  "Calculation failed")
```

Or use `expect_snapshot()`, which we'll discuss shortly. `expect_error()` also comes with variants `expect_warning()` and `expect_message()` for testing for warnings and messages in the same way as errors. These are rarely needed for testing Shiny apps but are very useful for testing packages.

User Interface Functions

You can use the same basic idea to test functions that you've extracted out of your UI code. But these require a new expectation, because manually typing out all the HTML would be tedious, so instead we use a snapshot test.[3] A snapshot expectation differs from other expectations primarily in that the expected result is stored in a separate snapshot file rather than in the code itself. Snapshot tests are most useful when you are designing complex user interface design systems, which is outside of the scope of most apps. So here I'll briefly show you the key ideas and then point you to additional resources to learn more.

Take this UI function we defined earlier:

```
sliderInput01 <- function(id) {
  sliderInput(id, label = id, min = 0, max = 1, value = 0.5, step = 0.1)
}

cat(as.character(sliderInput01("x")))
#> <div class="form-group shiny-input-container">
#>   <label class="control-label" id="x-label" for="x">x</label>
#>   <input class="js-range-slider" id="x" data-skin="shiny" data-min="0"
#>     data-max="1" data-from="0.5" data-step="0.1" data-grid="true"
#>     data-grid-num="10" data-grid-snap="false" data-prettify-separator=","
#>     data-prettify-enabled="true" data-keyboard="true" data-data-type="number"/>
#> </div>
```

How would we test that this output is as we expect? We could use `expect_equal()`:

```
test_that("shinyInput01() creates expected HTML", {
  expect_equal(
    as.character(sliderInput01("x")),
    "<div class=\"form-group shiny-input-container\">\n
      <label class=\"control-label\" id=\"x-label\" for=\"x\">x</label>\n
      <input class=\"js-range-slider\" id=\"x\" data-skin=\"shiny\" data-min=\"0\"
        data-max=\"1\" data-from=\"0.5\" data-step=\"0.1\" data-grid=\"true\"
        data-grid-num=\"10\" data-grid-snap=\"false\" data-prettify-separator=\",\"
```

---

[3] Snapshot tests require the third edition of testthat. New packages will automatically use the testthat 3e (*https://testthat.r-lib.org/articles/third-edition.html*), but you'll need to manually update older packages.

```
          data-prettify-enabled=\"true\" data-keyboard=\"true\"
          data-data-type=\"number\"/>\n
      </div>"
    )
  })
  #> Test passed 🎊
```

But the presence of quotes and newlines requires a lot of escaping in the string—that makes it hard to see exactly what we expect and, if the output changes, makes it hard to see exactly what's happened.

The key idea of snapshot tests is to store the expected results in a separate file: that keeps bulky data out of your test code and means that you don't need to worry about escaping special values in a string. Here we use expect_snapshot() to capture the output displayed on the console:

```
test_that("shinyInput01() creates expected HTML", {
  expect_snapshot(sliderInput01("x"))
})
```

The main difference with other expectations is that there's no second argument that describes what you expect to see. Instead, that data is saved in a separate *.md* file. If your code is in *R/slider.R* and your test is in *tests/testthat/test-slider.R*, then snapshot will be saved in *tests/testhat/_snaps/slider.md*. The first time you run the test, expect_snapshot() will automatically create the reference output, which will look like this:

```
# shinyInput01() creates expected HTML

Code
  sliderInput01("x")
Output
  <div class="form-group shiny-input-container">
    <label class="control-label" id="x-label" for="x">x</label>
    <input class="js-range-slider" id="x" data-skin="shiny" data-min="0"
      data-max="1" data-from="0.5" data-step="0.1" data-grid="true"
      data-grid-num="10" data-grid-snap="false" data-prettify-separator=","
      data-prettify-enabled="true" data-keyboard="true" data-data-type="number"/>
  </div>
```

If the output later changes, the test will fail. You either need to fix the bug that causes it to fail or, if it's a deliberate change, update the snapshot by running test that::snapshot_accept().

It's worth contemplating the output here before committing to this as a test. What are you really testing here? If you look at how the inputs become the outputs, you'll notice that most of the output is generated by Shiny and only a very small amount is the result of your code. That suggests this test isn't particularly useful: if this output changes, it's much more likely to be the result of a change to Shiny than the result of a change to your code. This makes the test fragile; if it fails, it's unlikely to be your fault, and fixing the failure is unlikely to be within your control.

You can learn more about snapshot tests in the "Snapshot tests" (*https://oreil.ly/GpHlU*) testthat vignette.

Workflow

Before we talk about testing functions that use reactivity or JavaScript, we'll take a brief digression to work on your workflow.

Code Coverage

It's very useful to verify that your tests test what you think they're testing. A great way to do this is with "code coverage," which runs your tests and tracks every line of code that is run. You can then look at the results to see which lines of your code are never touched by a test, and this gives you the opportunity to reflect on if you've tested the most important, highest-risk, or hardest-to-program parts of your code. It's not a substitute for thinking about your code—you can have 100% test coverage and still have bugs. But it's a fun and a useful tool to help you think about what's important, particularly when you have complex nested code.

I won't cover it in detail here, but I highly recommend trying it out with `devtools::test_coverage()` or `devtools::test_coverage_file()`. The main thing to notice is that green lines are tested, while red lines are not.

Code coverage supports a slightly different workflow:

1. Use `test_coverage()` or `test_coverage_file()` to see which lines of code are tested.

2. Look at untested lines and design tests specifically to test them.

3. Repeat until all important lines of code are tested. (Getting to 100% test coverage often isn't worth it, but you should check that you are hitting the most critical parts of your app.)

Code coverage also works with the tools for testing reactivity and (to some extent) JavaScript, so it's a useful foundational skill.

Keyboard Shortcuts

If you followed the advice in "Benefits" on page 292, then you can already run tests just by typing `test()` or `test_file()` at the console. But tests are something that you'll do so often it's worth having a keyboard shortcut at your fingertips. RStudio has one useful shortcut built in: Cmd/Ctrl+Shift+T runs `devtools::test()`. I recommend that you add three to complete the set:

- Bind Cmd/Ctrl+T to `devtools::test_file()`
- Bind Cmd/Ctrl+Shift+R to `devtools::test_coverage()`
- Bind Cmd/Ctrl+R to `devtools::test_coverage_file()`

You're of course free to choose whatever shortcut makes sense to you, but these share some underlying structure. Keyboard shortcuts using Shift apply to the whole package, and shortcuts without Shift apply to the current file.

Figure 21-1 shows what my keyboard shortcuts look like on a Mac.

Report test coverage for a file	Cmd+R	Addin
Report test coverage for a package	Shift+Cmd+R	Addin
Run a test file	Cmd+T	Addin

Figure 21-1. My keyboard shortcut for a Mac.

Workflow Summary

Here's a summary of all the techniques I've talked about so far:

- From the R file, use `usethis::use_test()` to create the test file (the first time it's run) or navigate to the test file (if it already exists).
- Write code/write tests. Press Cmd/Ctrl+T to run the tests and review the results in the console. Iterate as needed.
- If you encounter a new bug, start by capturing the bad behavior in a test. In the course of making the minimal code, you'll often get a better understanding of where the bug lies, and having the test will ensure that you can't fool yourself into thinking that you've fixed the bug when you haven't.
- Press Cmd/Ctrl+R to check that you're testing what you think you're testing.
- Press Cmd/Ctrl+Shift+T to make sure you haven't accidentally broken anything else.

Testing Reactivity

Now that you understand how to test regular nonreactive code, it's time to move on to challenges specific to Shiny. The first challenge is testing reactivity. As you've already seen, you can't run reactive code interactively:

```
x <- reactive(input$y + input$z)
x()
#> Error: Operation not allowed without an active reactive context.
```

```
#> * You tried to do something that can only be done from inside a reactive
#> consumer.
```

You might wonder about using reactiveConsole() like we did in Chapter 15. Unfortunately, its simulation of reactivity depends fundamentally on an interactive console, so it won't work in tests.

Not only does the reactive error when we attempt to evaluate it, even if it did work, input$y and input$z wouldn't be defined. To see how it works, let's start with a simple app that has three inputs, one output, and three reactives:

```r
ui <- fluidPage(
  numericInput("x", "x", 0),
  numericInput("y", "y", 1),
  numericInput("z", "z", 2),
  textOutput("out")
)
server <- function(input, output, session) {
  xy <- reactive(input$x - input$y)
  yz <- reactive(input$z + input$y)
  xyz <- reactive(xy() * yz())
  output$out <- renderText(paste0("Result: ", xyz()))
}
```

To test this code we'll use the testServer(). This function takes two arguments: a server function and some code to run. The code is run in a special environment, *inside* the server function, so you can access outputs, reactives, and a special session object that allows you to simulate user interaction. The main time you'll use this is for session$setInputs(), which allows you to set the value of input controls, as if you were a user interacting with the app in a browser:

```r
testServer(server, {
  session$setInputs(x = 1, y = 1, z = 1)
  print(xy())
  print(output$out)
})
#> [1] 0
#> [1] "Result: 0"
```

(You can abuse testServer() to get in an interactive environment that does support reactivity: testServer(myApp(), browser()).)

Note that we're only testing the server function; the ui component of the app is completely ignored. You can see this most clearly by inspecting the inputs: unlike a real Shiny app, all inputs start as NULL, because the initial value is recorded in the ui. We'll come back to UI testing in "Testing JavaScript" on page 309.

```r
testServer(server, {
  print(input$x)
})
#> NULL
```

Now that you have a way to run code in a reactive environment, you can combine it with what you already know about testing code to create something like this:

```
test_that("reactives and output updates", {
  testServer(server, {
    session$setInputs(x = 1, y = 1, z = 1)
    expect_equal(xy(), 0)
    expect_equal(yz(), 2)
    expect_equal(output$out, "Result: 0")
  })
})
#> Test passed 🎉
```

Once you've mastered the use of `testServer()`, then testing reactive code becomes almost as easy as testing nonreactive code. The main challenge is debugging failing tests: you can't step through them line by line like a regular test, so you'll need to add a `browser()` inside of `testServer()` so that you can interactively experiment to diagnose the problem.

Modules

You can test a module in a way that's similar to testing an app function, but here it's a little more clear that you're only testing the server side of the module. Let's start with a simple module that uses three outputs to display a brief summary of a variable:

```
summaryUI <- function(id) {
  tagList(
    outputText(ns(id, "min")),
    outputText(ns(id, "mean")),
    outputText(ns(id, "max")),
  )
}
summaryServer <- function(id, var) {
  stopifnot(is.reactive(var))

  moduleServer(id, function(input, output, session) {
    range_val <- reactive(range(var(), na.rm = TRUE))
    output$min <- renderText(range_val()[[1]])
    output$max <- renderText(range_val()[[2]])
    output$mean <- renderText(mean(var()))
  })
}
```

We'll use `testServer()` as we did previously, but the call is a little different. As before, the first argument is the server function (now the module server), but now we also need to supply additional arguments in a list called `args`. This takes a list of arguments to the module server (the `id` argument is optional; `testServer()` will fill it in automatically if omitted). Then we finish up with the code to run:

```
x <- reactiveVal(1:10)
testServer(summaryServer, args = list(var = x), {
  print(range_val())
  print(output$min)
})
#> [1]  1 10
#> [1] "1"
```

Again, we can turn this into an automated test by putting it inside test_that() and calling some expect_ functions. Here I wrap it all up into a test that checks that the module responds correctly as the reactive input changes:

```
test_that("output updates when reactive input changes", {
  x <- reactiveVal()
  testServer(summaryServer, args = list(var = x), {
    x(1:10)
    session$flushReact()
    expect_equal(range_val(), c(1, 10))
    expect_equal(output$mean, "5.5")

    x(10:20)
    session$flushReact()
    expect_equal(range_val(), c(10, 20))
    expect_equal(output$min, "10")
  })
})
#> Test passed 🎉
```

There's one important trick here: because x is created outside of testServer(), changing x does not automatically update the reactive graph, so we have to do so manually by calling session$flushReact().

If your module has a return value (a reactive or list of reactives), you can capture it with session$getReturned(). Then you can check the value of that reactive, just like any other reactive:

```
datasetServer <- function(id) {
  moduleServer(id, function(input, output, session) {
    reactive(get(input$dataset, "package:datasets"))
  })
}

test_that("can find dataset", {
  testServer(datasetServer, {
    dataset <- session$getReturned()

    session$setInputs(dataset = "mtcars")
    expect_equal(dataset(), mtcars)

    session$setInputs(dataset = "iris")
    expect_equal(dataset(), iris)
  })
```

```
})
#> Test passed 🍃
```

Do we need to test what happens if `input$dataset` isn't a dataset? In this case, we don't because we know that the module UI restricts the options to valid choices. That's not obvious from inspection of the server function alone.

Limitations

`testServer()` is a simulation of your app. The simulation is useful because it lets you quickly test reactive code, but it is not complete.

- Unlike the real world, time does not advance automatically. So if you want to test code that relies on `reactiveTimer()` or `invalidateLater()`, you'll need to manually advance time by calling `session$elapse(millis = 300)`.
- `testServer()` ignores UI. That means inputs don't get default values, and no JavaScript works. Most importantly this means that you can't test the `update*` functions, because they work by sending JavaScript to the browser to simulate user interactions. You'll require the next technique to test such code.

Testing JavaScript

`testServer()` is only a limited simulation of the full Shiny app, so any code that relies on a "real" browser running will not work. Most importantly, this means that no JavaScript will be run. This might not seem important because we haven't talked about JavaScript in this book, but there are a number of important Shiny functions that use it behind the scenes:

- All `update*()` functions: "Updating Inputs" on page 153.
- `showNotification()`/`removeNotification()`: "Notifications" on page 126.
- `showModal()`/`hideModal()`: "Explicit Confirmation" on page 136.
- `insertUI()`/`removeUI()`/`appendTab()`/`insertTab()`/`removeTab()`: we'll cover this later in the book.

To test these functions, you need to run the Shiny app in a real browser. You could, of course, do this yourself using `runApp()` and clicking around, but we want to automate that process so that you run your tests frequently. We'll do this with an off-label use of the shinytest (*https://rstudio.github.io/shinytest*) package. You can use shinytest as the website recommends, automatically generating test code using an app, but since you're already familiar with testthat, we'll take a different approach, constructing tests by hand.

We'll work with one R6 object from the shinytest package: `ShinyDriver`. Creating a new `ShinyDriver` instance starts a new R process that runs your Shiny app and a *headless* browser. A headless browser works just like a usual browser, but it doesn't have a window that you can interact with; the sole means of interaction is via code. The primary downsides of this technique is that it's slower than the other approaches (it takes at least a second for even the simplest apps), and you can only test the outside of the app (i.e., it's harder to see the values of reactive variables).

Basic Operation

To demonstrate the basic operation, I'll create a very simple app that greets you by name and provides a reset button:

```
ui <- fluidPage(
  textInput("name", "What's your name"),
  textOutput("greeting"),
  actionButton("reset", "Reset")
)
server <- function(input, output, session) {
  output$greeting <- renderText({
    req(input$name)
    paste0("Hi ", input$name)
  })
  observeEvent(input$reset, updateTextInput(session, "name", value = ""))
}
```

To use shinytest, you start an app with `app <- ShinyDriver$new()`, interact with it using `app$setInputs()` and friends, then get values returned by `app$getValue()`:

```
app <- shinytest::ShinyDriver$new(shinyApp(ui, server))
app$setInputs(name = "Hadley")
app$getValue("greeting")
#> [1] "Hi Hadley"
app$click("reset")
app$getValue("greeting")
#> [1] ""
```

Every use of shinytest begins by creating a ShinyDriver object with `ShinyDriver$new()`, which takes a Shiny app object or a path to a Shiny app. It returns an R6 object that you interact with much like the session object you encountered previously, using `app$setInputs()`: it takes a set of name-value pairs, updates the controls in the browser, and then waits until all reactive updates are complete.

The first difference is that you'll need to explicitly retrieve values using `app$getValue(name)`. Unlike with `testServer()`, you can't access the values of reactives using ShinyDriver because it can only see what a user of the app can see. But there's a special Shiny function called `exportTestValues()` that creates a special output that shinytest can see but a human cannot.

There are two other methods that allow you to simulate other actions:

- `app$click(name)` clicks a button called `name`.

- `app$sendKeys(name, keys)` sends key presses to an input control called `name`. `keys` will normally be a string like `app$sendKeys(id, "Hi!")`. But you can also send special keys using `webdriver::key`, à la `app$sendKeys(id, c(webdriver::key$control, "x"))`. Note that any modifier keys will be applied to all subsequent key presses, so you'll need multiple calls if you want some key presses with modifiers and some without.

See `?ShinyDriver` for more details and a list of more esoteric methods.

As before, once you've figured out the appropriate sequence of actions interactively, you can turn it into a test by wrapping in `test_that()` and calling expectations:

```
test_that("can set and reset name", {
  app <- shinytest::ShinyDriver$new(shinyApp(ui, server))
  app$setInputs(name = "Hadley")
  expect_equal(app$getValue("greeting"), "Hi Hadley")

  app$click("reset")
  expect_equal(app$getValue("greeting"), "")
})
```

The background Shiny app and web browser are automatically shut down when the app object is deleted and collected by the garbage collector. If you're not familiar with what that means, you might find the section "Unbinding and the Garbage Collector" from *Advanced R* (*https://oreil.ly/zXYN9*) helpful.

Case Study

We'll finish up with a case study exploring how you might test a more realistic example, combining both `testServer()` and shinytest. We'll use a radio-button control that also provides a free-text "other" option. This might look familiar, as we used it before as a motivation for developing a module in "Limited Selection and Other" on page 275:

```
ui <- fluidPage(
  radioButtons("fruit", "What's your favourite fruit?",
    choiceNames = list(
      "apple",
      "pear",
      textInput("other", label = NULL, placeholder = "Other")
    ),
    choiceValues = c("apple", "pear", "other")
  ),
  textOutput("value")
)
```

```
server <- function(input, output, session) {
  observeEvent(input$other, ignoreInit = TRUE, {
    updateRadioButtons(session, "fruit", selected = "other")
  })

  output$value <- renderText({
    if (input$fruit == "other") {
      req(input$other)
      input$other
    } else {
      input$fruit
    }
  })
}
```

The actual computation is quite simple. We *could* consider pulling the `renderText()`
expression out into its own function:

```
other_value <- function(fruit, other) {
  if (fruit == "other") {
    other
  } else {
    fruit
  }
}
```

But I don't think it's worth it because the logic here is very simple and not generaliza-
ble to other situations. I think the net effect of pulling this code out of the app into a
separate file would make the code harder to read.

So we'll start by testing the basic flow of reactivity: do we get the correct value after
setting `fruit` to an existing option? And do we get the correct value after setting
`fruit` to other and adding some free text?

```
test_that("returns other value when primary is other", {
  testServer(server, {
    session$setInputs(fruit = "apple")
    expect_equal(output$value, "apple")

    session$setInputs(fruit = "other", other = "orange")
    expect_equal(output$value, "orange")
  })
})
#> Test passed 🎉
```

That doesn't check that other is automatically selected when we start typing in the
other box. Unfortunately, we can't test that using `testServer()` because it relies on
`updateRadioButtons()`:

```
test_that("returns other value when primary is other", {
  testServer(server, {
```

```
    session$setInputs(fruit = "apple", other = "orange")
    expect_equal(output$value, "orange")
  })
})
#> — Failure (<text>:2:3): returns other value when primary is other ————————
#> output$value (`actual`) not equal to "orange" (`expected`).
#>
#> `actual`:   "apple"
#> `expected`: "orange"
#> Backtrace:
#>   1. shiny::testServer(...)
#>  22. testthat::expect_equal(output$value, "orange")
```

So now we need to use ShinyDriver:

```
test_that("automatically switches to other", {
  app <- ShinyDriver$new(shinyApp(ui, server))
  app$setInputs(other = "orange")
  expect_equal(app$getValue("fruit"), "other")
  expect_equal(app$getValue("value"), "orange")
})
```

Generally, you are best off using `testServer()` as much as possible and only using `ShinyDriver` for the bits that need a real browser.

Testing Visuals

What about components like plots or HTML widgets where it's difficult to describe the correct appearance using code? You can use the final, richest, and most fragile testing technique: save a screenshot of the affected component. This combines screenshotting from shinytest with whole-file snapshotting from testthat. It works similarly to the snapshotting described in "User Interface Functions" on page 302, but instead of saving text into an *.md* file, it creates a *.png* file. This also means that there's no way to see the differences on the console, so you'll instead be prompted to run `testthat::snapshot_review()`, which uses a Shiny app to visualize the differences.

The primary downside of testing using screenshots is that even the tiniest of changes requires a human to confirm that it's OK. This is a problem because it's hard to get different computers to generate pixel-reproducible screenshots. Differences in operating system, browser version, and even font versions can lead to screenshots that look the same to a human but are very slightly different. This generally means that visual tests are best run by one person on their local computer, and it's generally not worthwhile to run them in a continuous integration tool. It is possible to work around these issues, but it's a considerable challenge and beyond the scope of this book.

Screenshotting individual elements in shinytest and whole-file snapshotting in testthat are both very new features, and it's still not clear to us what the ideal interface is. So for now, you'll need to string the pieces together yourself, using code like this:

```
path <- tempfile()
app <- ShinyDriver$new(shinyApp(ui, server))

# Save screenshot to temporary file
app$takeScreenshot(path, "plot")
#
expect_snapshot_file(path, "plot-init.png")

app$setValue(x = 2)
app$takeScreenshot(path, "plot")
expect_snapshot_file(path, "plot-update.png")
```

The second argument to `expect_snapshot_file()` gives the filename that the image will be saved in a file snapshot directory. If these tests are in a file called *test-app.R*, then these two file snapshots will be saved in *tests/testthat/_snaps/app/plot-init.png* and *tests/testthat/_snaps/app/plot-update.png*. You want to keep the names of these files short but evocative enough to remind you what you're testing if something goes wrong.

Philosophy

This chapter has focused mostly on the mechanics of testing, which are most important when you get started with testing. But you'll soon get the mechanics under your belt, and your questions will become more structural and philosophical.

I think it's useful to think about false positives and false negatives: it's possible to write tests that don't fail when they should and do fail when they shouldn't. I think when you start testing, your biggest struggles are with false positives: how do you make sure your tests are actually catching bad behavior? But I think you move past this fairly quickly.

When Should You Write Tests?

When should you write tests? There are three basic options:

Before you write the code
> This is a style of code called test-driven development, and if you know exactly how a function should behave, it makes sense to capture that knowledge as code *before* you start writing the implementation.

After you write the code

While writing code, you'll often build up a mental to-do list of worries about your code. After you've written the function, turn these into tests so that you can be confident that the function works the way that you expect.

When you start writing tests, beware writing them too soon. If your function is still actively evolving, keeping your tests up to date with all the changes is going to feel frustrating. That may indicate you need to wait a little longer.

When you find a bug

Whenever you find a bug, it's good practice to turn it into an automated test case. This has two advantages. First, to make a good test case, you'll need to relentlessly simplify the problem until you have a very minimal reprex that you can include in a test. Second, you'll make sure that the bug never comes back again!

Summary

This chapter has shown you how to organize your app into a package so that you can take advantage of the powerful tools provided by the testthat package. If you've never made a package before, this can seem overwhelming, but as you've seen, a package is just a simple set of conventions that you can readily adapt for a Shiny app. This requires a little up-front work but unlocks a big payoff: the ability to automate tests radically increases your ability to write complex apps.

In the next chapter, you'll learn out how to figure out what's making your apps slow and some techniques for making them faster.

## CHAPTER 22

Security

Most Shiny apps are deployed within a company firewall, and since you can generally assume that your colleagues aren't going to try and hack your app,[1] you don't need to think about security. If, however, your app contains data that only some of your colleagues should be able to access, or you want to expose your app to the public, you will need to spend some time on security. When securing your app, there are two main things to protect:

- Your data: you want to make sure an attacker can't access any sensitive data.

- Your compute resources: you want to make sure an attacker can't mine bitcoin or use your server as part of a spam farm.

Fortunately your job is made a little easier because security is a team sport. Whoever deploys your app is responsible for security *between* apps, ensuring that app A can't access the code or data in app B, and can't steal all the memory and compute power on the server. Your responsibility is the security *within* your app, making sure that an attacker can't abuse your app to achieve their ends. This chapter will give the basics of securing your Shiny, broken down into securing your data and securing your compute resources.

If you're interested in learning a little more about security and R in general, I highly recommend "R and Security" (*https://oreil.ly/BM159*), Colin Gillespie's entertaining and educational useR! 2019 talk. Let's begin by loading shiny:

```
library(shiny)
```

---

[1] If you can't assume that, you have bigger problems! That said, some companies do have a "zero-trust" model, so you should double-check with your IT team.

Data

The most sensitive data is stuff like personally identifying information (PII), regulated data, credit card data, health data, or anything else that would be a legal nightmare for your company if it was made public. Fortunately, most Shiny apps don't deal with those types of data,[2] but there is an important type of data you do need to worry about: passwords. You should never include passwords in the source code of your app. Instead, either put them in environment variables or, if you have many, use the config (*https://github.com/rstudio/config*) package. Either way, make sure that they are never included in your source code control by adding the appropriate files to .gitignore. I also recommend documenting how a new developer can get the appropriate credentials.

Alternatively, you may have data that is user-specific. If you need to *authenticate* users (i.e., identify them through a user name and password), never attempt to roll a solution yourself. There are just too many things that might go wrong. Instead, you'll need to work with your IT team to design a secure access mechanism. You can see some best practices in the "Kerberos with RStudio Pro Products" (*https://oreil.ly/zixmG*) and "Securing Deployed Content" (*https://oreil.ly/tsrAJ*) RStudio documentation. Note that code within server() is isolated, so there's no way for one user session to see data from another. The only exception is if you use caching—see "Cache Scope" on page 338 for details.

Finally, note that Shiny inputs use client-side validation—that is, the checks for valid input are performed by JavaScript in the browser, not by R. This means it's possible for a knowledgeable attacker to send values that you don't expect. For example, take this simple app:

```
secrets <- list(
  a = "my name",
  b = "my birthday",
  c = "my social security number",
  d = "my credit card"
)

allowed <- c("a", "b")
ui <- fluidPage(
  selectInput("x", "x", choices = allowed),
  textOutput("secret")
)
server <- function(input, output, session) {
  output$secret <- renderText({
    secrets[[input$x]]
```

---

2 If your app does work these types of data, it's imperative that you partner with a software engineer with security expertise.

```
    })
  }
```

You might expect that a user could access my name and birthday but not my social security number or credit card details. But a knowledgeable attacker can open up a JavaScript console in their browser and run `Shiny.setInputValue("x", "c")` to see my SSN. So to be safe, you need to check all user inputs from your R code:

```
server <- function(input, output, session) {
  output$y <- renderText({
    req(secrets$x %in% allowed)
    secrets$y[[secrets$x == input$x]]
  })
}
```

I deliberately didn't create a user-friendly error message—the only time you'd see it would be if you're trying to break the app, and we don't need to help out an attacker.

Compute Resources

It's hopefully obvious that the following app is very dangerous, because it allows the user to run any R code they want. They could delete important files, modify data, or send confidential data back to the user of the app:

```
ui <- fluidPage(
  textInput("code", "Enter code here"),
  textOutput("results")
)
server <- function(input, output, session) {
  output$results <- renderText({
    eval(parse(text = input$code))
  })
}
```

In general, the combination of `parse()` and `eval()` is a big warning sign for any Shiny app:[3] they instantly make your app vulnerable. Similarly, you should never `source()` an uploaded *.R* file or `rmarkdown::render()` an uploaded *.Rmd*. But these cases are pretty obvious and are unlikely to be the source of real problems.

The bigger challenge arises because there are a number of functions that `parse()`, `eval()`, or both in a way that you're not aware of. Here are the most common:

---

3 The only exception is if they don't involve user-supplied data in any way.

Model formulas

It's possible to construct a model that executes arbitrary R code:

```
df <- data.frame(x = 1:5, y = runif(5))
mod <- lm(y ~ {print("Hi!"); x}, data = df)
#> [1] "Hi!"
```

This makes it difficult to safely allow a user to define their own models.

Glue labels

The glue package provides a powerful way to create strings from data:

```
title <- "foo"
number <- 1
glue::glue("{title}-{number}")
#> foo-1
```

But `glue()` evaluates anything inside of `{}`:

```
glue::glue("{title}-{print('Hi'); number}")
#> [1] "Hi"
#> foo-1
```

If you want to allow a user to supply a glue string to generate a label, instead use `glue::glue_safe()`, which only looks up variable names and doesn't evaluate code:

```
glue::glue_safe("{title}-{number}")
#> foo-1
glue::glue_safe("{title}-{print('Hi'); number}")
#> Error in .transformer(expr, env): object 'print('Hi'); number' not found
```

Variable transformation

There's no way to safely allow a user to provide code snippets to transform a variable for dplyr or ggplot2. You might expect they'll write `log10(x)`, but they could write `{print("Hi"); log10(x)}`.

This also means that you should never use the older `ggplot2::aes_string()` with user supplied input. Instead, stick with the techniques in Chapter 12.

The same problem can occur with SQL. For example, if you construct SQL with `paste()`, for example:

```
find_student <- function(name) {
  paste0("SELECT * FROM Students WHERE name = ('", name, "');")
}
find_student("Hadley")
#> [1] "SELECT * FROM Students WHERE name = ('Hadley');"
```

an attacker can provide a malicious username:[4]

```
find_student("Robert'); DROP TABLE Students; --")
#> [1] "SELECT * FROM Students WHERE name = ('Robert'); DROP TABLE Students; --');"
```

This looks a bit odd, but it's a valid SQL query in three parts:

- `SELECT * FROM Students WHERE name = ('Robert');` finds a student with the name Robert.

- `DROP TABLE Students;` deletes the `Students` table (!!).

- `--'` is a comment needed to prevent the extra `'` from turning into a syntax error.

To avoid this problem, never generate SQL strings with paste, and instead use a system that automatically escapes user input (like dbplyr (*https://dbplyr.tidyverse.org*)), or use `glue::glue_sql()`:

```
con <- DBI::dbConnect(RSQLite::SQLite(), ":memory:")
find_student <- function(name) {
  glue::glue_sql("SELECT * FROM Students WHERE name = ({name});", .con = con)
}
find_student("Robert'); DROP TABLE Students; --")
#> <SQL> SELECT * FROM Students WHERE name = ('Robert''); DROP TABLE Students; --');
```

It's a little hard to tell at first glance, but this is safe, because SQL's equivalent of `\'` is `''`, so the query returns all rows of the `Students` table where the name is literally "Robert'); DROP TABLE Students; –".

---

4 This example was inspired by Little Bobby Tables (*https://xkcd.com/327*).

## CHAPTER 23

Performance

A Shiny app can support thousands or tens of thousands of users, if developed the right way. But most Shiny apps are quickly thrown together to solve a pressing analytic need and typically begin life with poor performance. This is a feature of Shiny: its allows you to quickly prototype a proof of concept that works for you, before figuring out how to make it fast so many people can use it simultaneously. Fortunately, it's generally straightforward to get 10–100x performance with a few simple tweaks. This chapter will show you how.

We'll begin with a metaphor: thinking about a Shiny app like a restaurant. Next, you'll learn how to *benchmark* your app, using the shinyloadtest package to simulate many people using your app at the same time. This is the place to start, because it lets you figure out you have a problem and helps measure the impact of any changes that you make.

Then you'll learn how to *profile* your app using the profvis package to identify slow parts of your R code. Profiling lets you see exactly where your code is spending its time, so you can focus your efforts where they're most impactful.

Finally, you'll learn a handful of useful techniques to *optimize* your code, improving the performance where needed. You'll learn how to cache reactives, how to move data prep code out of your app, and how to use a little applied psychology to help your app *feel* as fast as possible.

I recommend watching Joe Cheng's rstudio::conf(2019) keynote "Shiny in Production: Principles, Practices, and Tools" (*https://oreil.ly/Yv3NA*) for a demo of the whole process of benchmarking, profiling, and optimizing. In that talk (and accompanying case study (*https://oreil.ly/C6d1e*)), Joe walks through the complete process with a realistic app.

```r
library(shiny)
```

Special thanks to my RStudio colleagues Joe Cheng, Sean Lopp, and Alan Dipert, whose RStudio::conf() talks were particularly helpful when writing this chapter.

Dining at Restaurant Shiny

When considering performance, it's useful to think of a Shiny app as a restaurant.[1] Each customer (user) comes into the restaurant (the server) and makes an order (a request), which is then prepared by a chef (the R process). This metaphor is useful because, like a restaurant, one R process can serve multiple users at the same time, and there are similar ways to deal with increasing demand.

To begin, you might investigate ways to make your current chef more efficient (optimize your R code). To do so, you'd first spend some time watching your chef work to find the bottlenecks in their method (profiling) and then brainstorming ways to help them work faster (optimizing). For example, maybe you could hire a prep cook who can come in before the first customer and chop some vegetables (preparing the data), or you could invest in a time-saving gadget (a faster R package).

Or you might think about adding more chefs (processes) to the restaurant (server). Fortunately, it's much easier to add more processes[2] than it is to hire trained chefs. If you keep hiring more chefs, eventually the kitchen (server) will get too full, and you'll need to add more equipment (memory of cores). Adding more resources to allow a server to run more processes is called scaling *up*.[3]

At some point, you'll have crammed as many chefs into your restaurant as you possibly can, and it's still not enough to meet demand. At that point, you'll need to build more restaurants. This is called scaling *out*,[4] and for Shiny it means using multiple servers. Scaling out allows you to handle any number of customers, as long as you can pay the infrastructure costs. I won't talk more about scaling out in this chapter, because while the details are straightforward, they depend entirely on your deployment infrastructure.

There's one major place where the metaphor breaks down: a normal chef can make multiple dishes at the same time, carefully interweaving the steps to take advantage of downtime in one recipe to work on another. R, however, is single-threaded, which

---

1 Thanks to Sean Lopp for this analogy from his rstudio::conf(2018) talk "Scaling Shiny: 10,000 User App" (*https://oreil.ly/0Ncsg*). I highly recommend watching it if you have any doubt that Shiny apps can handle thousands of users.

2 Again, this depends on exactly how your app is deployed, but typically you can dynamically control the number of processes based on the number of users. See "Scaling and Performance Tuning in RStudio Connect" (*https://oreil.ly/c4CGT*) for advice on RStudio's deployment offerings.

3 Or vertical scaling.

4 Or horizontal scaling.

means that it can't do multiple things at the same time. This is fine if all of the meals are fast to cook, but if someone requests a 24-hour sous vide pork belly, all later customers will have to wait 24 hours before the chef can start on their meal. Fortunately, you can work around this limitation using async programming (*https://rstudio.github.io/promises*), a complex topic that is beyond the scope of this book.

Benchmark

You almost always start by developing an app for yourself: your app is a personal chef who only ever has to serve one customer at a time (you!). While you might be happy with their performance right now, you might also worry that they won't be able to handle the 10 folks who need to use your app at the same time. Benchmarking lets you check the performance of your app with multiple users, without actually exposing real people to a potentially slow app. Or if you want to serve hundreds or thousands of users, benchmarking will help you figure out just how many users each process can handle and hence how many servers you'll need to use.

The benchmarking process is supported by the shinyloadtest (*https://rstudio.github.io/shinyloadtest*) package and has three basic steps:

1. Use `shinyloadtest::record_session()` to record a script simulating a typical user.
2. Replay the script with multiple simultaneous users with the shinycannon command-line tool.
3. Call `shinyloadtest::report()` to analyze the results.

Here I'll give an overview of how each of the steps work; if you need more details, check out shinyloadtest's documentation and vignettes.

Recording

If you're benchmarking on your laptop, you'll need to use two different R processes—one for Shiny and one for shinyloadtest.[5]

- In the first process, start your app and copy the URL that it gives you:

    ```
    runApp("myapp.R")
    #> Listening on http://127.0.0.1:7716
    ```

- In the second process, paste the URL into a `record_session()` call:

---

5 The easiest way to do this with RStudio is to open another RStudio instance. Alternatively, open a terminal and type R.

```
shinyloadtest::record_session("http://127.0.0.1:7716")
```

`record_session()` will open a new window containing a version of your app that records everything you do with it. Now you need to interact with the app to simulate a "typical" user. I recommend starting with a written script to guide your actions—this will make it easier to repeat in the future, if you discover there's some important piece missing. Your benchmarking will only be as good as your simulation, so you'll need to spend some time thinking about how to simulate a realistic interaction with the app. For example, don't forget to add pauses to reflect the thinking time that a real user would need.

Once you're done, close the app, and shinyloadtest will save *recording.log* to your working directory. This records every action in a way that can easily be replayed. Keep a hold of it as you'll need it for the next step.

(While benchmarking works great on your laptop, you likely want to simulate the eventual deployment as closely as possible in order to get the most accurate results. So if your company has a special way of serving Shiny apps, talk to your IT folks about setting up an environment that you can use for load testing.)

Replay

Now you have a script that represents the actions of a single user, and we'll next use it to simulate many people using a special tool called shinycannon. shinycannon is a bit of extra work to install because it's not an R package. It's written in Java because the Java language is particularly well suited to the problem of performing tens or hundreds of web requests in parallel, using as few computational resources as possible. This makes it possible for your laptop to both run the app and simulate many users. So start by installing shinycannon, following these instructions on RStudio (*https://oreil.ly/NkKZ2*).

Then run shinycannon from the terminal with a command like this:

```
shinycannon recording.log http://127.0.0.1:7911 \
  --workers 10 \
  --loaded-duration-minutes 5 \
  --output-dir run1
```

There are six arguments to `shinycannon`:

- The first argument is a path to the recording that you created in the previous step.

- The second argument is the URL to your Shiny app (which you copied and pasted in the previous step).

- `--workers` sets the number of parallel users to simulate. The previous command will simulate the performance of your app as if 10 people were using it simultaneously.

- `--loaded-duration-minutes` determines how long to run the test for. If this is longer than your script takes, shinycannon will start the script again from the beginning.

- `--output-dir` gives the name of the directory to save the output. You're likely to run the load test multiple times as you experiment with performance improvements, so strive to give informative names to these directories.

When load testing for the first time, it's a good idea to start with a small number of workers and a short duration in order to quickly spot any major problems.

Analysis

Now that you've simulated your app with multiple users, it's time to look at the results. First, load the data into R using `load_runs()`:

```r
library(shinyloadtest)
df <- load_runs("scaling-testing/run1")
```

This produces a tidy tibble that you can analyze by hand if you want. But typically you'll create the standard shinyloadtest report. This is an HTML report that contains the graphical summaries that the Shiny team has found to be most useful:

```r
shinyloadtest_report(df, "report.html")
```

I'm not going to discuss all the pages in the report here. Instead I'll focus on what I think is the most important plot: the session duration. To learn more about the other pages, I highly recommend reading the Analyzing Load Test Logs (*https://oreil.ly/VYyxh*) article.

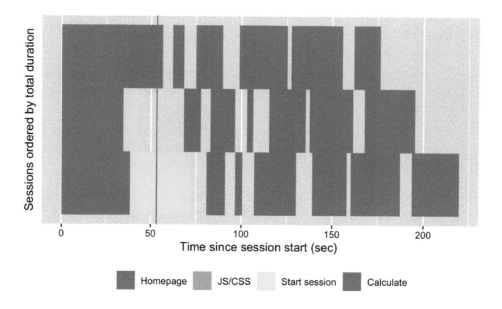

The *session duration* plot displays each simulated user session as a row. Each event is a rectangle with width proportional to time taken, colored by the event type. The red line shows the time that the original recording took.

When looking at this plot, consider the following questions:

- Does the app perform equivalently under load as it does for a single user? If so, congratulations! Your app is already fast enough, and you can stop reading this chapter ☺. Is the slowness in the "Homepage"? If so, you're probably using a ui function, and you're accidentally doing too much work there.

- Is "Start session" slow? That suggests the execution of your server function is slow. Generally, running the server function should be fast because all you're doing is defining the reactive graph (which is run in the next step). If it's slow, move expensive code either outside of server() (so it's run once on app startup) or into a reactive (so it's run on demand).

  Otherwise, and most typically, the slowness will be in "Calculate," which indicates that some computation in your reactive is slow, and you'll need to use the techniques in the rest of the chapter to find and fix the bottlenecks.

Profiling

If your app is spending a lot of time calculating, you next need to figure out which calculation is slow—that is, you need to *profile* your code to find the bottleneck. We're going to do profiling with the profvis (*https://rstudio.github.io/profvis*) package, which provides an interactive visualization of the profiling data collected by `utils::Rprof()`. I'll start by introducing the flame graph, the visualization used for profiling, then show you how to use profvis to profile R code and Shiny apps.

The Flame Graph

Across programming languages, the most common tool used to visualize profiling data is the *flame graph*. To help you understand it, I'm going to start by revisiting the basics of code execution, then build up progressively to the final visualization.

To make the process concrete, we'll work with the following code, where I use `pause()` (more on that shortly) to indicate work being done:

```
library(profvis)

f <- function() {
  pause(0.2)
  g()
  h()
  10
}
g <- function() {
  pause(0.1)
  h()
}
h <- function() {
  pause(0.3)
}
```

If I asked you to mentally run `f()` then explain what functions were called, you might say something like this:

- We start with `f()`.
- Then `f()` calls `g()`.
- Then `g()` calls `h()`.
- Then `f()` calls `h()`.

This is a bit hard to follow because we can't see exactly how the calls are nested, so instead you might adopt a more conceptual description:

- f
- f > g
- f > g > h
- f > h

Here we've recorded a list of call stacks, which you might remember from "Reading Tracebacks" on page 71, when we talked about debugging. The call stack is just the complete sequence of calls leading up to a function.

We could convert that list to a diagram by drawing a rectangle around each function name:

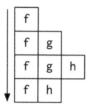

I think it's most natural to think about time flowing downward, from top to bottom, in the same way you usually think about code running. But by convention, flame graphs are drawn with time flowing from left to right, so we rotate our diagram by 90 degrees:

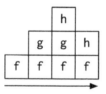

We can make this diagram more informative by making the width of each call proportional to the amount of time it takes. I also added some grid lines in the background to make it easier to check my work:

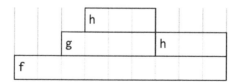

Finally, we can clean it up a little by combining adjacent calls to the same function:

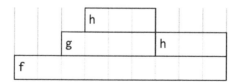

This is a flame graph! It's easy to see both how long f() takes to run and why it takes that long (i.e., where its time is spent).

You might wonder why it's a called a flame graph. Most flame graphs in the wild are randomly colored with "warm" colors, meant to evoke the idea of the computer running "hot." However, since those colors don't add any additional information, we usually omit them and stick to black and white. You can learn more about this color scheme, alternatives, and the history of flame graphs in "The Flame Graph" (*https://oreil.ly/AKXNP*).

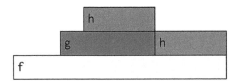

Profiling R Code

Now that you understand the flame graph, let's apply it to real code with the profvis package. It's easy to use: just wrap the code you want to profile in profvis::prof vis():

```
profvis::profvis(f())
```

After the code has completed, profvis will pop up an interactive visualization, as shown in Figure 23-1. You'll notice that it looks very similar to the graphs that I drew by hand, but the timings aren't exactly the same. That's because R's profiler works by stopping execution every 10 ms and recording the call stack. Unfortunately, we can't always stop at exactly the time we want because R might be in the middle of something that can't be interrupted. This means that the results are subject to a small amount of random variation; if you reprofiled this code, you'd get another slightly different result.

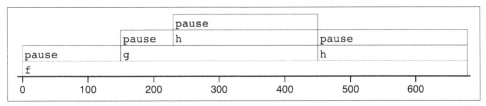

Figure 23-1. Results of profiling f() with profvis. X-axis shows elapsed time in ms, and y-axis shows depth of call stack.

As well as a flame graph, profvis also does its best to find and display the underlying source code so that you can click on a function in the flame graph to see exactly what's run.

Profiling a Shiny App

Not much changes when profiling a Shiny app. To see the difference, I'll make a very simple app that wraps around `f()`:

```
ui <- fluidPage(
  actionButton("x", "Push me"),
  textOutput("y")
)
server <- function(input, output, session) {
  output$y <- eventReactive(input$x, f())
}

# Note the explicit call to runApp() here: this is important
# as otherwise the app won't actually run.
profvis::profvis(runApp(shinyApp(ui, server)))
```

The results are shown in Figure 23-2.

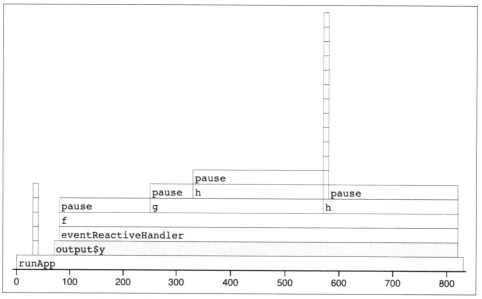

Figure 23-2. Results of profiling a Shiny app that uses f(). Note that the call stack is deeper and we have a couple of tall towers.

The output looks very similar to the last run. There are a couple of differences:

- `f()` is no longer at the bottom of the call stack. It's now on the fourth level because it's called by `eventReactiveHandler()` (the internal function that powers `eventReactive()`), which is triggered by output$y, which is wrapped inside `runApp()`.

- There are two very tall towers. Generally, these can be ignored because they don't take up much time and will vary from run to run because of the stochastic nature of the sampler. If you do want to learn more about them, you can hover to find out the function calls. In this case, the short tower on the left is the setup of the `eventReactive()` call, and the tall tower on the right is R's byte code compiler being triggered.

For more details, I recommend the profvis documentation, particularly its FAQs (*https://rstudio.github.io/profvis/faq.html*).

Limitations

The most important limitation of profiling is due to the way it works: R has to stop the process and inspect what R functions are currently run. That means that R has to be in control. There are a few places where this doesn't happen:

- Certain C functions don't regularly check for user interruptions. These are the same C functions you can't use Esc/Ctrl+C to cancel. That's generally not a good programming practice, but they do exist in the wild.

- `Sys.sleep()` asks the operating system to "park" the process for some amount of time, so R is not actually running. This is why we had to use `profvis::pause()` previously.

- Downloading data from the internet is usually done in a different process, so it won't be tracked by R.

Improve Performance

The most efficient way to improve performance is to find the slowest thing in the profile and try to speed it up. Once you've isolated a slow part, make sure it's wrapped in a standalone function (Chapter 18). Then make a minimal snippet of code that re-creates the slowness, reprofiling it to check that you captured it correctly. You'll rerun this snippet multiple times as you try out possible improvements. I also recommend writing a few tests (Chapter 21), because in my experience the easiest way to make code faster is to make it incorrect 😆.

Shiny code is just R code, so most techniques for improving performance are general. Two good places to start are the "Improving Performance" (*https://oreil.ly/JKsXC*) section of *Advanced R* and *Efficient R Programming* (*https://oreil.ly/LBGxs*) by Colin Gillespie and Robin Lovelace. I'm not going to repeat their advice here: instead, I'll focus on the techniques that are most likely to affect your Shiny app. I also highly recommend Alan Dipert's rstudio::conf(2018) talk "Make Shiny Fast by Doing as Little Work as Possible" (*https://oreil.ly/8wknl*).

Begin by resolving any issues where existing code is run more often than you expect. Make sure you're not repeating the same work in multiple reactives and that the reactive graph isn't updating more often than you expect ("The Reactlog Package" on page 220).

Next, I'll discuss the easiest way to improve the performance of your app, using caching to remember and replay slow calculations. I'll finish up with two other techniques that can help many Shiny apps: pulling out expensive preprocessing into a separate step and carefully managing user expectations

Caching

Caching is a very powerful technique for improving code performance. The basic idea is to record the inputs to and outputs from every call to a function. When the cache function is called with a set of inputs that it's already seen, it can replay the recorded output without recomputing. Packages like memoise (*https://memoise.r-lib.org*) provide tools for caching regular R functions.

Caching is particularly effective for Shiny apps, because the cache can be shared across users. That means when many people are using the same app, only the first user needs to wait for the results to be computed, then everyone else gets a speedy result from the cache.

Shiny provides a general tool for caching any reactive expression or render function: `bindCache()`.[6] As you know, reactive expressions already cache the most recently computed value; `bindCache()` allows you to cache any number of values and to share those values across users. I'll introduce you to the basics of `bindCache()`, show you a couple of practical examples, and then talk through some of the details of the cache "key" and scope. I recommend starting with "Using Caching in Shiny to Maximize Performance" (*https://oreil.ly/qLHab*) and "Using bindCache() to Speed Up an App" (*https://oreil.ly/guiyx*) if you want to learn more.

Basics

`bindCache()` is easy to use. Simply pipe either the `reactive()` or `render*` function that you want to cache into `bindCache()`:

```
r <- reactive(slow_function(input$x, input$y)) %>%
  bindCache(input$x, input$y)

output$text <- renderText(slow_function2(input$z)) %>%
  bindCache(input$z)
```

---

6 This function was introduced in Shiny 1.6.0, generalizing the older `renderCachedPlot()`, which only worked for plots.

The additional arguments are the cache keys: these are the values that are used to determine if a computation has been seen before. We'll discuss the cache keys in more details after showing a couple of practical uses.

Caching a Reactive

A common place to use caching is in conjunction with a web API: even if the API is very quick, you still have to send the request, wait for the server to respond, and then parse the result. So caching API results often yields a big performance improvement. Let's illustrate that with a simple example using the gh (*https://gh.r-lib.org*) package that talks to GitHub's API.

Imagine you want to design an app that shows what people have been working on lately. Here I've written a little function that gets the data from GitHub's event API and does some simple rectangling (*https://oreil.ly/iPTxi*) to turn it into a tibble:

```r
library(purrr)

latest_events <- function(username) {
  json <- gh::gh("/users/{username}/events/public", username = username)
  tibble::tibble(
    repo = json %>% map_chr(c("repo", "name")),
    type = json %>% map_chr("type"),
  )
}

system.time(hadley <- latest_events("hadley"))
#>    user  system elapsed
#>   0.138   0.033   0.743
head(hadley)
#> # A tibble: 6 x 2
#>   repo                     type
#>   <chr>                    <chr>
#> 1 hadley/r4ds              IssuesEvent
#> 2 hadley/mastering-shiny   IssuesEvent
#> 3 hadley/mastering-shiny   IssueCommentEvent
#> 4 hadley/mastering-shiny   IssuesEvent
#> 5 hadley/mastering-shiny   IssueCommentEvent
#> 6 hadley/mastering-shiny   IssuesEvent
```

And I can turn that into a very simple app:

```r
ui <- fluidPage(
  textInput("username", "GitHub user name"),
  tableOutput("events")
)
server <- function(input, output, session) {
  events <- reactive({
    req(input$username)
    latest_events(input$username)
  })
```

```
  output$events <- renderTable(events())
}
```

This app is going to feel a little sluggish because every time you type in a username, it's going to have to re-request the data, even if you just asked for it 15 seconds ago. We can dramatically improve performance by using bindCache():

```
server <- function(input, output, session) {
  events <- reactive({
    req(input$username)
    latest_events(input$username)
  }) %>% bindCache(input$username)
  output$events <- renderTable(events())
}
```

You might have spotted a problem with this approach—what happens if you come back to it tomorrow and request data for the same user? You'll get today's data, even though there might have been new activity. There's an implicit dependency on time that you need to make explicit. You can do that by adding Sys.Date() to the cache key so that the cache effectively only lasts for a single day:

```
server <- function(input, output, session) {
  events <- reactive({
    req(input$username)
    latest_events(input$username)
  }) %>% bindCache(input$username, Sys.Date())
  output$events <- renderTable(events())
}
```

You might worry that the cache will steadily accumulate data from past days that you'll never look at again, but fortunately the cache has a fixed total size and is smart enough to automatically remove the least-recently-used data when it needs more space.

Caching Plots

Most of the time you'll cache reactives, but you can also use bindCache() with render functions. Most render functions are pretty speedy, but there's one that can be slow if you have complex graphics: renderPlot().

For example, take the following app. If you run it yourself, you'll notice that the first time you show each plot, it takes a noticeable fraction of a second to render because it has to draw ~50,000 points. But the *next* time you draw each plot, it appears instantly because it's retrieved from the cache:

```
library(ggplot2)

ui <- fluidPage(
  selectInput("x", "X", choices = names(diamonds), selected = "carat"),
  selectInput("y", "Y", choices = names(diamonds), selected = "price"),
```

```
    plotOutput("diamonds")
)

server <- function(input, output, session) {
  output$diamonds <- renderPlot({
    ggplot(diamonds, aes(.data[[input$x]], .data[[input$y]])) +
      geom_point()
  }) %>% bindCache(input$x, input$y)
}
```

(If the .data syntax is unfamiliar to you, see Chapter 12 for details.)

There's one special consideration when it comes to caching plots: each plot is drawn in a variety of sizes, because the default plot occupies 100% of the available width, which varies as you resize the browser. That flexibility doesn't work very well for caching, because even a single pixel difference in the size would mean that the plot couldn't be retrieved from the cache. To avoid this problem, bindCache() caches plots with fixed sizes. The defaults are carefully chosen to "just work" in most cases, but if needed you can control with the sizePolicy argument and learn more in the ? sizeGrowthRatio.

Cache Key

It's worth talking briefly about the cache key: the set of values used to figure out whether or not the computation has been previously performed. These values are also used to determine the reactive dependencies, much like the *first* argument of observ eEvent() or eventReactive(). That means if you use the wrong cache key, you can get very confusing results. For example, imagine that I have this cached reactive:

```
r <- reactive(input$x + input$y) %>% bindCache(input$x)
```

If input$y changes, r() will not recompute. And if the result is retrieved from the cache, it will be the sum of the current value of x and whatever value y happened to have when the value was cached.

So the cache key should always include all of the reactive inputs in the expression. But you may also want to include additional values that are not used in the reactive. The most useful example of this is adding the current day, or some rounded current time, so that cached values are only used for a fixed amount of time.

As well as inputs, you can use other reactive()s as cache keys, but you'll need to keep them as a simple as possible (i.e., atomic vectors or simple lists of atomic vectors). Don't use large datasets because it is expensive to figure out if a large data frame has already been seen, and that will reduce the benefit you see from caching.

Cache Scope

By default, the plot cache is stored in memory, is never bigger than 200 MB, is shared across all users in a single process, and is lost when the app restarts. You can change this default for individual reactives or for the whole session:

- bindCache(…, cache = "session") will use a separate cache for each user session. This ensures that private data is not potentially shared between users, but it also reduces the benefit of caching.
- Use shinyOptions(cache = cachem::cache_mem()) or shinyOptions(cache = cachem::cache_disk()) to change the default cache across the whole app. You can use them to make sure a cache is shared across multiple processes and lasts across app restarts. See ?bindCache for more details.

It's also possible to chain multiple caches together or write your own custom storage backend. You can learn more about these options in the documentation for cachem (*https://cachem.r-lib.org*), the caching package that powers bindCache().

Other Optimizations

There are two other optimizations that crop up in many apps: performing data import and manipulation on a schedule and carefully managing user expectations.

Schedule Data Munging

Imagine that your Shiny app uses a dataset that requires a little initial data cleaning. The data prep is relatively complicated and takes a nontrivial amount of time. You've discovered that it's a bottleneck for your app and want to do better.

Let's pretend that you've already extracted the code out into a function, and it looks something like this:

```
my_data_prep <- function() {
  df <- read.csv("path/to/file.csv")
  df %>%
    filter(!not_important) %>%
    group_by(my_variable) %>%
    some_slow_function()
}
```

And currently you call it in your server function:

```
server <- function(input, output, session) {
  df <- my_data_prep()
  # Lots more code
}
```

The server function is called every time a new session starts, but the data is always the same, so you can immediately make your app faster (and use less memory) by moving the data processing out of `server()`:

```
df <- my_data_prep()
server <- function(input, output, session) {
  # Lots more code
}
```

Since you're paying attention to this code, it's also worth checking that you're using the most efficient way to load your data:

- If you have a flat file, try `data.table::fread()` or `vroom::vroom()` instead of `read.csv()` or `read.table()`.
- If you have a data frame, try saving with `arrow::write_feather()` and reading with `arrow::read_feather()`. Feather is a binary file format that can be considerably faster[7] to read and write.
- If you have objects that aren't data frames, try using `qs::qread()`/`qs::qsave()` instead of `readRDS()`/`saveRDS()`.

If these changes aren't enough to resolve the bottleneck, you might consider using a separate cron job or scheduled RMarkdown report to call `my_data_prep()` and save the results. Then your app can load the pre-prepared data and get to work. This is like hiring a prep chef who comes in at three a.m. (when there are no customers) so that during the lunch rush your chefs can be as efficient as possible.

Manage User Expectations

Finally, there are a few tweaks you can make to your app design to make it feel faster and improve the overall user experience of your app. Here are four tips that can be used in many apps:

- Split your app up into tabs, using `tabsetPanel()`. Only outputs on the current tab are recomputed, so you can use this to focus computation on what the user is currently looking at.
- Require a button press to start a long-running operation. Once the operation starts, let the user know what's happening using the techniques of "Notifications" on page 126. If possible, display an incremental progress bar ("Progress Bars" on page 129) because there's good evidence that progress bars make operations feel faster (*https://oreil.ly/MG0G8*).

---

7 See this Ursa Labs blog post (*https://oreil.ly/Xtr73*) for some benchmarks.

- If the app requires significant work to happen on startup (and you can't reduce it with preprocessing), make sure to design your app so that the UI can still appear and you can let the user know that they'll need to wait.

- Finally, if you want to keep the app responsive while some expensive operation happens in the background, it's time to learn about async programming (*https://rstudio.github.io/promises/index.html*).

Summary

This chapter has given you the tools to precisely measure and improve the performance of any Shiny app. You learned about shinyloadtest to measure the performance and using shinycannon to simulate multiple users working with your app at the same time. Then you learned how to use profvis to find the single-most expensive operation and a grab bag of techniques that you can use to improve it.

This is the last chapter in *Mastering Shiny*—thank you for making it all the way to the end! I hope you have found the book useful and that the skills I have given you help you produce many compelling Shiny apps. I'd love to hear if you've found the book useful or if there's anything that you think could be improved in the future. The best way to get in touch is on Twitter, @hadleywickham (*https://twitter.com/hadleywickham*), or on GitHub (*https://github.com/hadley/mastering-shiny*). Thanks again for reading, and best wishes for your future Shiny apps!

Index

Symbols

%<-% operator, 273
<- (assignment operator), 7, 267

A

action buttons
    dialog box considerations, 136
    on click events, 46
    Other button, 275
    paired start and stop buttons, 239
    paired with observeEvent(), 20
    tweaking text of, 155
actionLink(), 20
Angular, 206
animations, pausing, 239
any_of(), 195
app.R, 3
assignment operator (<-), 7, 267
asynchronous programming, 325, 340
authentication, 318
automated bookmarking, 183

B

background color, 100
benchmarking, 325-328
best practices (see also module system; performance improvement; testing)
    code extraction, 251-257
    packages, 287-296
    security, 317-321
    software engineering, 245-250
bindCache(), 334
bookmarkButton(), 180
bookmarking, 179-184

Bootstrap framework, 97
browser(), 109
brush events, 105, 109
brushedPoints(), 109, 113
brushOpts(), 111
bslib package, 98-101

C

cache key, 337
caching, 334-338
case studies
    accidental injuries investigation
        accessing narratives from dataset, 63
        data exploration, 53
        dataset, 51
        practice exercises, 64
        prototype app, 57
        rate versus count, 61
        table aesthetics, 60
    debugging, 75-79
    file transfer, 149
    histograms, 271
    JavaScript testing, 311
    module system
        dynamic UI, 282-284
        overview of, 275
        wizard interface, 278-281
    reactive graphs, 237
    reprexes (reproducible examples), 83
    selecting numeric variables, 268
cheat sheet, 10
checkboxGroupInput(), 19
circular references, 160
class argument, 21

.click argument, 25, 105
client-side validation, 318
code (see also best practices)
    creating UIs with, 166-177
    extracting imperative, 37
    extracting into independent apps, 251-257
    isolating, 228
    organizing, 246
    reducing duplicated, 8, 42
    sharing across apps, 293
    source code management, 248
code coverage, 304
code examples, obtaining and using, xvii
code reviews, 249
column layout, 91
column(), 38, 93
comments and questions, xviii
compute resources, 319
conditions, 123
config package, 248, 318
confirming and undoing
    explicit confirmation, 136
    trash, 139
    undoing actions, 137
continuous integration/deployment (CI/CD),
    249
CSS frameworks, 97, 101
.csv (comma-separated values), 145
curly braces ({}), 23

D

data munging, 338
data-masking
    base R code, 194
    example: dplyr, 191
    example: ggplot2, 189
    tidy-selection and, 196
    user-supplied data, 193
    uses for, 187
data-variable, 186
datasets
    converting apps into packages, 288
    downloading, 145
    limiting to data frames or matrices, 267
    reloading uploaded data, 255
    uploading, 143
dataTableOutput(), 24
dateInput(), 18, 253
dateRangeInput(), 18

dblclick argument, 25, 105
debouncing, 31
debugging
    case study, 75-79
    interactive debugger, 74
    main cases of problems, 70
    reading tracebacks, 71
    reprexes, 80-87
    tracebacks in Shiny, 72
    unexpected reactive firing, 79
declarative programming, 31
dependency management, 247
deploying app-packages, 294
development workflow
    app creation, 68
    controlling the view, 70
    seeing your changes, 69
devtools::load_all(), 292
devtools::test_file(), 300
dialog boxes, 136, 175
distinct(), 196
downloadButton(), 25, 144
downloadHandler(), 144
downloadLink(), 25, 144
dplyr, 191
drop-down menus, 97
dynamic user interfaces (dynamic UI) (see also
    inputs; outputs)
    creating UI with code, 166-177
        dialog boxes, 175
        dynamic filtering, 171-175
        isolate(), 167
        multiple controls, 168
        practice exercises, 176
        uiOutput() and renderUI(), 166
    dynamic visibility, 162-166
        conditional UI, 163
        practice exercises, 166
        tabsetPanel(), 162
        wizard interface, 165
    module system, 282-284
    techniques for creating, 153
    updating inputs, 153-162
        circular references, 160
        freezing reactive inputs, 158
        hierarchical select boxes, 156
        interrelated inputs, 160
        practice exercises, 161
        simple uses, 155

update functions, 153
dynamism, 218

E

eagerness, 226
Ember, 206
env-variable, 186
error handling, 225
error messages, 120
eval(), 197, 319
evaluation, controlling timing of, 44-48 (see
    also tidy evaluation)
event-driven programming, 203
eventExpr argument, 49
eventReactive(), 48-49, 131, 229
execution order, 34
exercises
    accidental injuries investigation, 64
    app creation, 10
    bookmarking, 184
    creating UI with code, 176
    downloads, 26
    dynamic visibility, 166
    file transfer, 151
    inputs, 21
    isolating code, 230
    module system, 265, 274
    page layouts, 94
    reactive expressions, 226
    reactive graphs, 217, 240
    reactive programming, 35
    reactive values, 224
    themes, 101
    timed invalidation, 233
    updating inputs, 161
expectations, 297
expect_equal(), 300
expect_error(), 300
explicit confirmation, 136

F

false positive/negative tests, 314
feedback(), 120
file transfer
    case study, 149
    downloads, 25, 144-148
    practice exercises, 151
    uploads, 20, 141-144
fileInput(), 20, 122, 141

fill, 111
fillPage(), 90
find_vars(), 270
fixedPage(), 90
flame graphs, 329
fluidPage(), 6, 90, 93
fluidRow(), 38, 93, 254
fonts, 100
foreground color, 100
FRAN (functional reactive animation), 206
freezeReactiveValue(), 159
frequency polygons, 36
functional programming, 254
Functional Reactive Programming, 206
functions
    data-masking, 187-195
    feedback functions, 120
    internal functions, 256
    layout functions, 89-97
    versus reactive programming, 202
    render functions, 7, 22, 29
    roles of in Shiny apps, 251
    server functions, 255
    UI functions, 252-255
    update functions, 153

G

gender identification, 277
generating a bookmark, 181
getting help, 80
ggplot2, 189
gh package, 335
Git/GitHub, 248
glue package, 320
graphics
    dynamic height and width, 115
    images, 116
    interactivity, 105-115

H

handlerExpr argument, 49
headless browsers, 310
height, dynamic, 115
hierarchical select boxes, 156
holding cells, 139
hover argument, 25, 105
HTML conventions, 97, 101

I

images, 116
imperative programming, 31
incProgress(), 130
indirection problem, 185
infectious eagerness, 226
input argument, 28
inputId, 16
inputs
    accumulating, 238
    basic UI
        action buttons, 20
        common structure, 16
        dates, 18
        file uploads, 20
        free text, 16
        limited choices, 18, 275
        numeric inputs, 17
    input controls, 6
    module system, 266-275
    one output and multiple inputs, 237
    page layouts, 89-97
    in reactive graphs, 213, 215-218
    security issues for user inputs, 320
    transferring files, 141-144
    updating for dynamic UIs, 153-162
    validating, 120
interactive debugger, 74
interactivity
    brushing, 109
    clicking, 107
    limitations of, 115
    modifying plots, 111-114
    mouse events, 105
    multiple interaction types, 109
    plotOutput(), 105
internal functions, 256
invalidateLater(), 230, 239, 309
isolate(), 167, 228, 239

J

JavaScript
    Shiny functions using, 309
    testing, 309-313

K

keyboard shortcuts, 304
Knockout, 206

L

label parameter, 16
layouts
    layout functions, 6
    multipage, 94-97
    practice exercises, 94
    single-page, 89-94
    themes, 98-101
load_all(), 288
long-running computation, 231

M

mainPanel(), 91
make_ui(), 282
memoise, 334
menus, drop-down, 97
message argument, 131
message(), 79
Meteor, 206
MobX, 206
modalDialog(), 136, 175
modal_confirm, 137
model formulas, 320
module system
    basics of, 261-266
    benefits of, 259
    inputs and outputs, 266-275
    modules inside of modules, 270
    reactive graphs and, 42
    role in app construction, 259
    single object modules, 284
    testing, 307
moduleServer(), 285
mouse events, 105
multipage layouts, 94-97
multirow layout, 93

N

namespacing, 264
naming conventions, 265
navbarMenu(), 96
navbarPage(), 96-97
navlistPanel(), 96
nearPoints(), 107
notifications
    progressive, 129
    removing on completion, 128
    showNotification(), 126

transient, 127
NS(), 283
numeric variables, 268
numericInput(), 15, 17, 169

O

observe(), 227, 237
observeEvent()
    automatically switching controls with, 164
    capturing and saving results, 138
    combined with reactiveValues(), 237
    debugging with, 49
    versus eventReactive(), 229
    versus observe(), 227
    paired with action links and buttons, 21
    updates on mouse clicks, 111
    updating inputs with, 154-156
observers, 226
on.exit(), 225
onBookmark(), 184
onRestore(), 184
"Other" button, 275
output argument, 29
output$plot, 22
outputs
    basic UI
        downloads, 25, 144
        placeholders created by, 22
        plots, 25
        tables, 24
        text, 22
    module system, 266-275
    multiple outputs, 272
    versus observers, 226
    one output and multiple inputs, 237
    output controls, 6
    page layouts, 89-97
    in reactive graphs, 214-215

P

packages
    additional resources, 287
    benefits of using, 292
    converting existing apps, 288-292
    core idea of, 287
    deploying app-packages, 294
    R CMD check, 294
    structure overview, 287
page functions, 90

page layouts (see layouts)
parameterized RMarkdown documents, 146
parse(), 197, 319
passwords, 16, 318
performance improvement
    additional resources, 333
    approach to, 333
    benchmarking, 325-328
    caching, 334-338
    data munging, 338
    demo video, 323
    managing user expectations, 339
    overview of, 323
    profiling, 328-333
    restaurant analogy, 324
personally identifying information (PII), 318
photos, 116
plot themes, 100
plotly package, 115
plotOutput(), 25, 105
polling, 231
print debugging, 79
producers, 36
profiling, 328-333
profvis package, 328
progress bars
    challenges of, 130
    Shiny, 130
    spinners, 133
    waiter package, 132
prototyping, 57
purrr::map(), 283
purrr::pmap(), 254

Q

questions and comments, xviii

R

R CMD check, 294
R packages
    additional resources, 295
    installing, xvi
radioButtons(), 18, 275
React, 206
reactive contexts, 29
reactive dependencies, 33
reactive expressions
    creating, 9, 41
    error handling, 225

need for, 43
versus observers and outputs, 226
overview of, 33
in reactive graphs, 212-213
reducing duplication with, 8
role in app construction
    comparing datasets, 36
    dual activity of, 36
    exploring multiple simulations, 38
    reactive graphs, 40
    simplifying reactive graphs, 41
reactive graphs
automated bookmarking, 183
escaping constraints of
    antipatterns, 240
    benefits and drawbacks of, 235
    case studies, 237-240
    input invalidation, 235
overview of, 33
role in app construction
    drawing reactive graphs, 40
    simplifying reactive graphs, 41
step-by-step tour of
    dynamism, 218
    input changes, 215-218
    reactlog package, 220
    sample app, 209
    session flow, 211-215
reactive programming
benefits and drawbacks of, 235
brief history of, 206
building blocks of
    controlling timing of evaluation, 44-48
    isolating code, 228
    observers and outputs, 49, 226
    reactive expressions, 225
    reactive values, 223
debugging, 79
definition of term, 206
essence of reactivity, 8, 202
introduction to
    execution order, 34
    imperative versus declarative program-
      ming, 31
    key idea of reactive programming, 27
    lazy updates, 32
    practice exercises, 35
    reactive expressions, 33, 36-44
    reactive graphs, 33

server functions, 27-30
    simple example, 30
  need for, 202-205
reactive(), 41
reactiveConsole(), 306
reactivePoll(), 231
reactiveTimer(), 45, 309
reactiveVal(), 111, 223
reactiveValues(), 223, 237
ReactiveX, 206
reactlog package, 220
rectangling, 335
removeModal(), 137
removeNotification(), 128
render functions, 7, 22, 29
renderDataTable(), 24
renderImage(), 116
renderPlot(), 25, 115, 336
renderPrint(), 7, 23
renderTable(), 7, 24
renderText(), 23
renderUI(), 167, 168, 175
renv, 247
reports, downloading, 146
reprexes (reproducible examples)
    benefits of, 80
    case study, 83
    creating smallest possible, 82
    making, 81
    simple example of, 81
req(), 107, 121-124, 143
RMarkdown, 146
rsconnect package, 294
RStudio
    installing, xvi
    interactive debugger, 74
    keyboard shortcuts, 304
RStudio Community site, 80
RStudio Connect, 293
RStudio Gadgets, 293
rule of three, 42
Run App button, 4
rxtools, 235

S

screenshots, 313
security
    additional resources, 317
    authentication, 318

compute resources, 319
data, 318
main items to protect, 317
passwords, 16, 318
responsibilities for, 317
select boxes, 156
selectInput(), 6, 15, 18, 121
selection tools, 105
separate(), 195
server functions
    handling long reactives, 255
    output$plot, 22
    overview of, 7, 27-30
server-side bookmarking, 183
session duration, 328
session$elapse(), 309
session$flushReact(), 308
session$getReturned(), 308
session$setInputs(), 306
setBookmarkExclude(), 183
sharing apps, 293
sharing code across apps, 293
Shiny
    approach to learning, xiv
    basic reactivity (see also reactive program-
        ming)
        controlling timing of evaluation, 44-48
        key idea of reactive programming, 27
        observers, 49
        reactive expressions, 36-44
        reactive programming, 30-35
        server functions, 27-30
    basic UI
        extension packages, 15
        file transfer, 141
        inputs, 15-22
        outputs, 22-25
        practice exercises, 21, 26
    benefits of, xiii, 201, 235
    installing, 3
    introduction to
        adding behavior to apps, 7
        adding UI controls to apps, 6
        app development process, 6, 67-70
            (see also workflow)
        creating app directories and files, 3
        operation cheat sheet, 10
        practice exercises, 10
        reducing duplicated code, 8

running and stopping apps, 4
prerequisites to learning, xv
target audience, xiv
use cases, xiii (see also case studies)
shinyapps.io, 293
shinycannon, 326
shinycssloaders package, 135
shinyFeedback package, 120
shinyloadtest package, 325
shinySignals, 235
shinytest package, 309
showModal(), 137
showNotification(), 126
sidebarLayout(), 91, 93
sidebarPanel(), 91
single object modules, 284
single-page layouts, 89-94
sliderInput(), 15, 17
snapshot tests, 302, 313
software engineering (see also best practices)
    code organization, 246
    code reviews, 249
    continuous integration/deployment, 249
    dependency management, 247
    recommendations for improvement, 245
    source code management, 248
    testing, 247
source(), 288, 319
special conditions, 123
spinners, 133
spreadsheets, 206
stroke, 111

T

tableOutput(), 6, 24
tabPanel(), 94
tabsetPanel(), 94-97, 183
testing
    best practices, 247
    false positives and false negatives, 314
    of functions, 298-304
    hierarchy of, 298
    levels of, 297
    of JavaScript, 309-313
    overview of, 297
    purpose of, 297
    of reactivity, 305-309
    snapshot tests, 302
    timing of, 314

of visuals, 313
workflow for, 304
testServer(), 306-309
testthat package, 297
testthat::snapshot_accept(), 303
testthat::snapshot_review(), 313
textAreaInput(), 16
textInput(), 15-17, 121, 124, 153
textOutput(), 22, 169
themes, 98-101
tidy evaluation
benefits and drawbacks of, 185
data-masking, 187-195
indirection problem, 185
parse() and eval(), 197
tidy-selection, 195-197
timed invalidation, 45, 230-233
timer accuracy, 232
timestamps, 233
titlePanel(), 91
tracebacks, 71-73
transformations, variable, 320
trash, 139
tReactive(), 21
.tsv (tab-separated value), 145
two-column layout, 91

U

UI functions, 252-255
uiOutput(), 167, 169
undoing actions, 137
update functions, 153
updateSelectInput(), 159
updateSliderInput(), 153, 155
updateTextInput(), 238
uploads, 20, 141-144
user bookmark command, 182
user expectations, 339
user feedback
confirming and undoing, 136-139
notification, 126-129
progress bars, 129-135

validation, 119-126
useShinyFeedback(), 120
usethis, 293
usethis::use_data(mydataset), 288
usethis::use_description(), 288, 291, 293
usethis::use_package(), 294
usethis::use_rstudio(), 288
use_waitress(), 132

V

validation
canceling execution, 121
ensuring text properties with, 17
importance of in app construction, 119
req() and, 124
shinyFeedback package, 120
validate(), 125
variable transformation, 320
variables, 202, 268
verbatimTextOutput(), 6, 22
version-control systems, 248
visibility, dynamic, 162-166
VisiCalc, 206
visuals, testing, 313
Vue.js, 206

W

waiter package, 132, 135
width, dynamic, 115
withProgress(), 130
wizard interface, 165, 278-281
workflow
debugging, 70-80
development, 67-70
reprexes, 80-87
testing, 304
when using packages, 292

Z

zeallot, 273

About the Author

Hadley Wickham is chief scientist at RStudio, a winner of the 2019 COPSS Presidents' Award, and a member of the R Foundation. He builds computational and cognitive tools to make data science easier, faster, and more fun. His work includes packages for data science (like the tidyverse, which includes ggplot2, dplyr, and tidyr) and principled software development (roxygen2, testthat, and pkgdown). He's also a writer, educator, and speaker, promoting the use of R for data science. Learn more on his website (*http://hadley.nz*).

Colophon

The animal on the cover of *Mastering Shiny* is a kererū (*Hemiphaga novaeseelandiae*), also known as the New Zealand pigeon, as it is the only pigeon species native to the New Zealand mainland.

The kererū's feathers are bronze-tinted and its head and body are a lustrous purple-green. This pigeon's underbelly is white and the bill red, matching its eyes. They emit soft *coos*, and make a distinctive wing-beating sound when landing or taking off. Typically slender and active, the kererū gains body mass during fruit and mating seasons.

Because they are one of the few native birds large enough to swallow fruit whole, kererū play a vital role in dispersing native seeds throughout New Zealand. This pigeon has acquired a reputation for being the drunkest bird in New Zealand—plump kererū have been known to fall from trees after consuming fermented fruit.

Kererū are culturally significant to the Māori people. For example, in Māori lore, it is said that when the trickster Māui searched the underworld for his parents, he took the form of a kererū. Though Māori would traditionally use the bird's meat, bones, and feathers, current preservation regulations put restrictions on hunting kererū.

Due to invasive species, hunting, and habitat deterioration, the New Zealand pigeon is listed as Near Threatened by the IUCN. Many of the animals on O'Reilly covers are endangered; all of them are important to the world.

Color illustration by Karen Montgomery, based on a black and white engraving from *British Birds*. The cover fonts are Gilroy Semibold and Guardian Sans. The text font is Adobe Minion Pro; the heading font is Adobe Myriad Condensed; and the code font is Dalton Maag's Ubuntu Mono.

O'REILLY®

There's much more where this came from.

Experience books, videos, live online training courses, and more from O'Reilly and our 200+ partners—all in one place.

Learn more at oreilly.com/online-learning